H. Endo · R. Frey (Eds.)

Anatomical Imaging

Towards a New Morphology

H. Endo · R. Frey (Eds.)

Anatomical Imaging

Towards a New Morphology

Springer

Hideki Endo, Ph.D.
Professor, The University Museum, The University of Tokyo
7-3-1 Hongo, Bunkyo-ku, Tokyo 113-0033, Japan

Roland Frey, Ph.D.
Research Group: Reproduction Management, Leibniz Institute for Zoo and Wildlife Research (IZW)
Alfred-Kowalke-Strasse 17, 10315 Berlin, Germany

ISBN 978-4-431-76932-3 Springer Tokyo Berlin Heidelberg New York
ISBN 978-4-431-76933-0 (eBook)

Library of Congress Control Number: 2008928682

Printed on acid-free paper

Springer is a part of Springer Science+Business Media
springer.com

Typesetting: SNP Best-set Typesetter Ltd., Hong Kong

Preface

From early on, anatomical imaging proceeded along three tracks that promoted each other mutually. One track consisted of human beings interested in the organic structure of animals and humans. This track comprises Paleo- and Neolithic hunters, shamans and clairvoyants of Sumeran and Babylonian times, dissecting artists of the Middle Ages, e.g., Leonardo da Vinci, and leads us to skilled and creative scientific anatomists of modern times. The second one consisted of technologies involved in the processing of images. Thus, a short walk through the history of anatomical imaging will bring us from cave paintings via written information only, via drawings, paintings, engravings, lithographies and prints to black and white photography, roentgenography, analogue colour photography, digital photography, offset printing, analogue and digital video, ultrasound imaging, magnetic resonance imaging, computed tomographic scanning and software-based three-dimensional reconstructions. The third one consisted of the media available or specifically made as a carrier for the images. It started with rock surfaces, then clay tablets, continued with parchment and paper, wooden blocks, copper plates, lime stones and brought us to photographic film, roentgen film and computer screens (Tab. 1).

Ever since the beginnings, presentation of the three-dimensional results of anatomical dissections had to choose between either the reduction to two-dimensional illustrations on a flat surface or the making of three-dimensional anatomical models. Both ways have been pursued for millennia. Two-dimensional figures predominated owing to much easier production and handling. This way was particularly successful since the availability of paper and even more so after invention of the printing press. However, sculptures (and bas-relief frescoes, etc) have long been a popular three-dimensional alternative for portraying anatomy. More recently, three-dimensional anatomical wax models were en vogue in the 18th century. Produced by a famous Italian school in Florence, more than one thousand were ordered by the Austrian emperor and transported on mules across the Alps to Vienna. They were used for anatomic teaching of military physicians and later for public education. The preservation of dissected specimens in formalin (a hydrate of formaldehyde) or similar substances has been a mainstay of anatomists not earlier than the second half of the nineteenth century. Most recently, a German anatomist invented the plastination of real bodies. Subsequent to dissection such methods, that substitute body fluid by synthetic resins, allow for directly presenting the intrinsic three-dimensional structure of an anatomical specimen.

Since the time of Vesalius printed books have been used to teach anatomy and to make anatomical results accessible to a greater audience. In most cases the translation from three to two dimensions was done by a skilled artist under guidance of the anatomist. The illustrations to Vesalius's epochal first comprehensive human anatomy text (1543) were made by Stephan van Kalkar, a student of Tizian. One of the latest textbooks on human anatomy (2005) applies a digital drawing technique to generate 'computed drawings' of almost 3-dimensional plasticity. This is part of the transition to modern computed virtual three-dimensional reconstructions of investigated specimens where three dimensions are provided by a machine. Serial scanned slices are reconstructed to create virtual translucent anatomical specimens. With these technologies the ancient dream of the classical anatomists came true: the look through the skin onto anatomical structures and the investigation of sections in any desired plane without destruction of the specimen. In addition to this new quality of visual investigations, computerized X-ray-based scanning transfers all data points of the real specimen automatically into an exactly corresponding virtual 3-dimensional space so that diverse quantifications of specific features can be executed on the computer screen.

At the moment modern technologies cannot replace anatomical dissections owing to a lack of resolution of the soft parts. Accordingly, best results can be obtained by using traditional methods and modern technologies in combination.

Against the above roughly sketched backdrop of the historical development of anatomical imaging, the aim of this book is not historical. Instead, it has been designed to provide an insight into modern anatomical imaging by presenting selected works of contemporary evolutionary morphologists, most of whom met July 31 – August 5 2005 at the IXth International Mammalogical Congress in Sapporo, Hokkaido, Japan. Chapters are arranged in correspondence with major parts of the vertebrate body (head, limbs, skeleton, reproductive organs, etc.). A short preface to each chapter explains the specific kind of anatomical imaging involved therein. Inevitably, such an insight provided by few authors, working in different fields and applying a variety of technologies, must remain fragmentary. Nonetheless we are confident to having made a book that can be an inspiration for other workers in the field, for academic and non-academic readers and for all those who are fascinated by the aesthetic appeal of anatomical structures.

Hideki Endo
Roland Frey

Contents

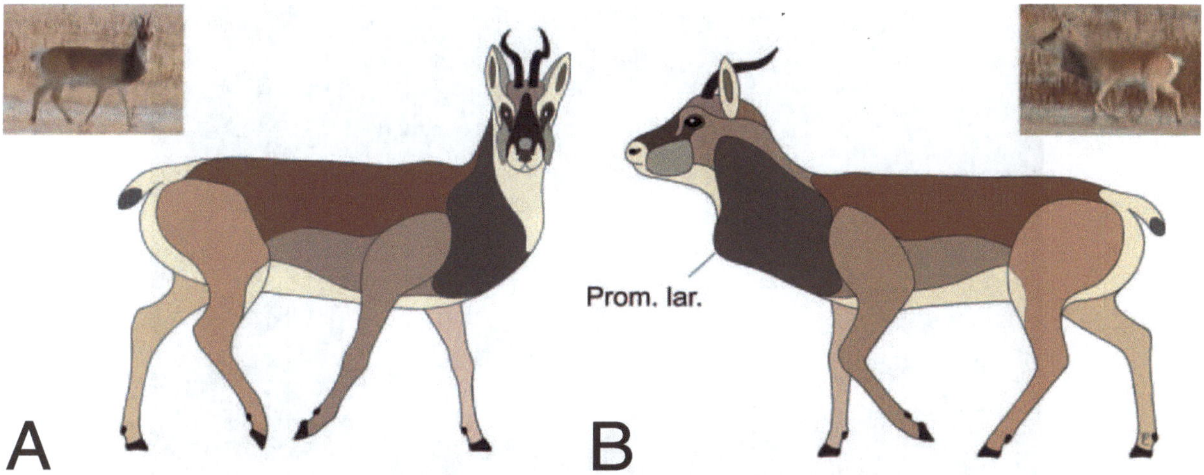

Fig. 1.2. Habitus and diagrammatic colour pattern of rutting male Mongolian gazelle according to video single frames (see insets). Laterofrontal view (**A**) and lateral view (**B**) evidencing the laryngeal prominence and the accentuation of the ventral neck region.

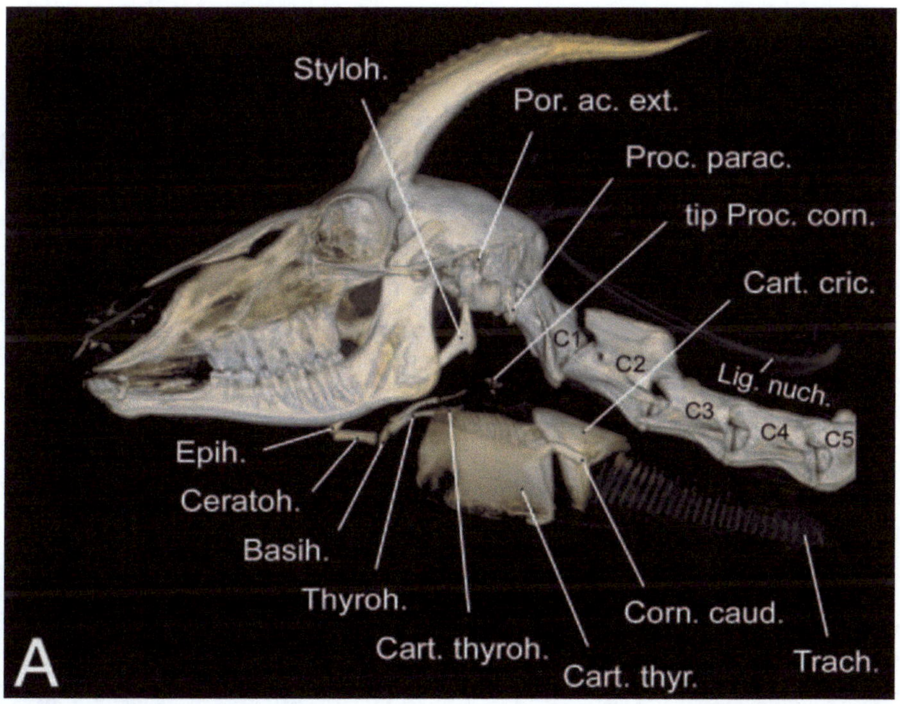

Fig. 1.3. Sexual dimorphism of the laryngeal cartilages as revealed by CT-based 3D imaging. Male (**A**) and female (**B**). (**A** alterated after Frey and Gebler 2003)

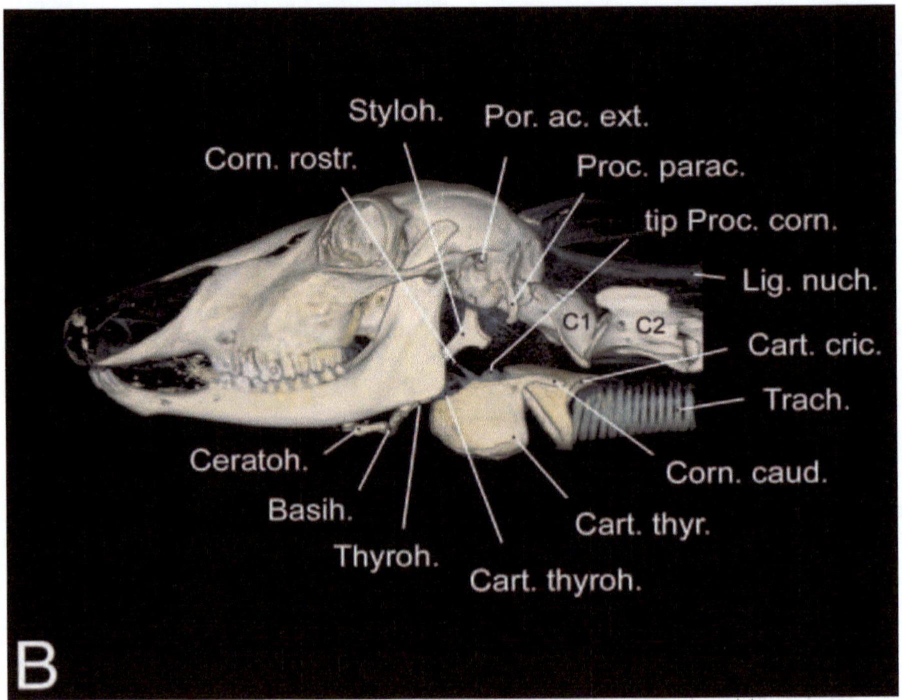

FIG. 1.3. *Continued*

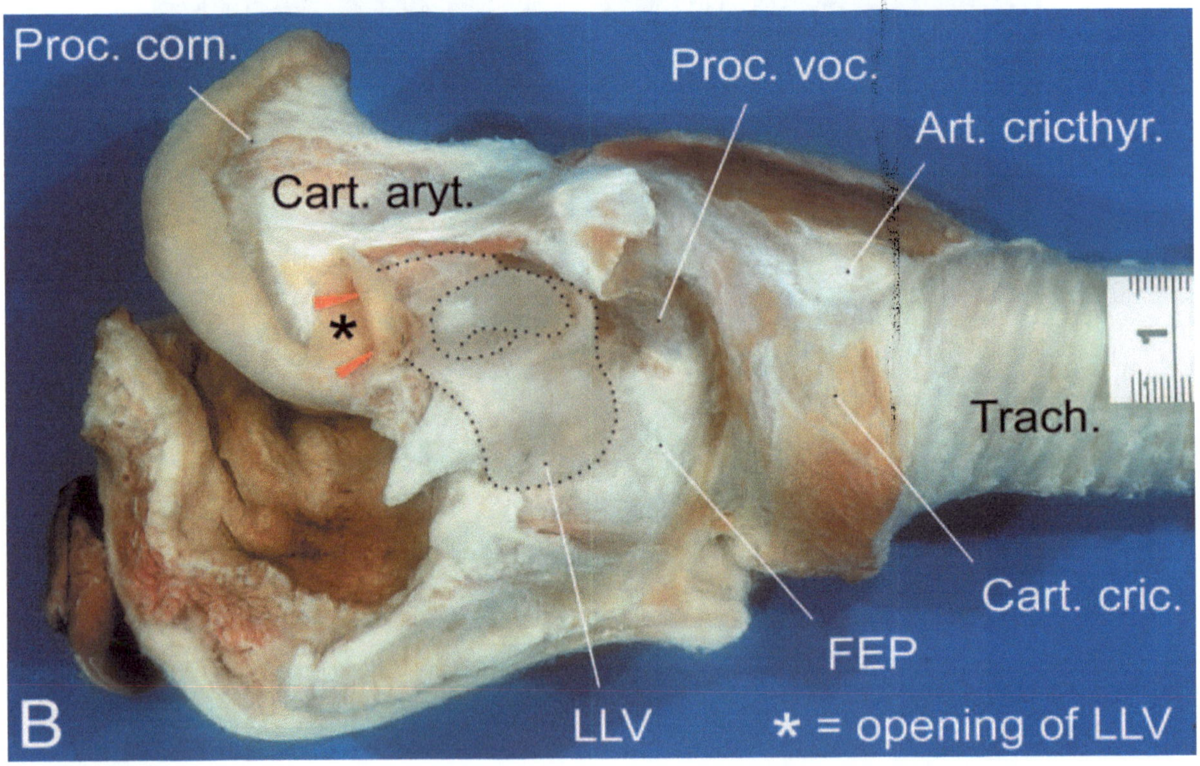

FIG. 1.4. Dissected left LLV of a male consisting of a crescent-shaped main chamber and a laterally protruding subsidiary sac (**B**). (**B** alterated after Frey and Riede 2003)

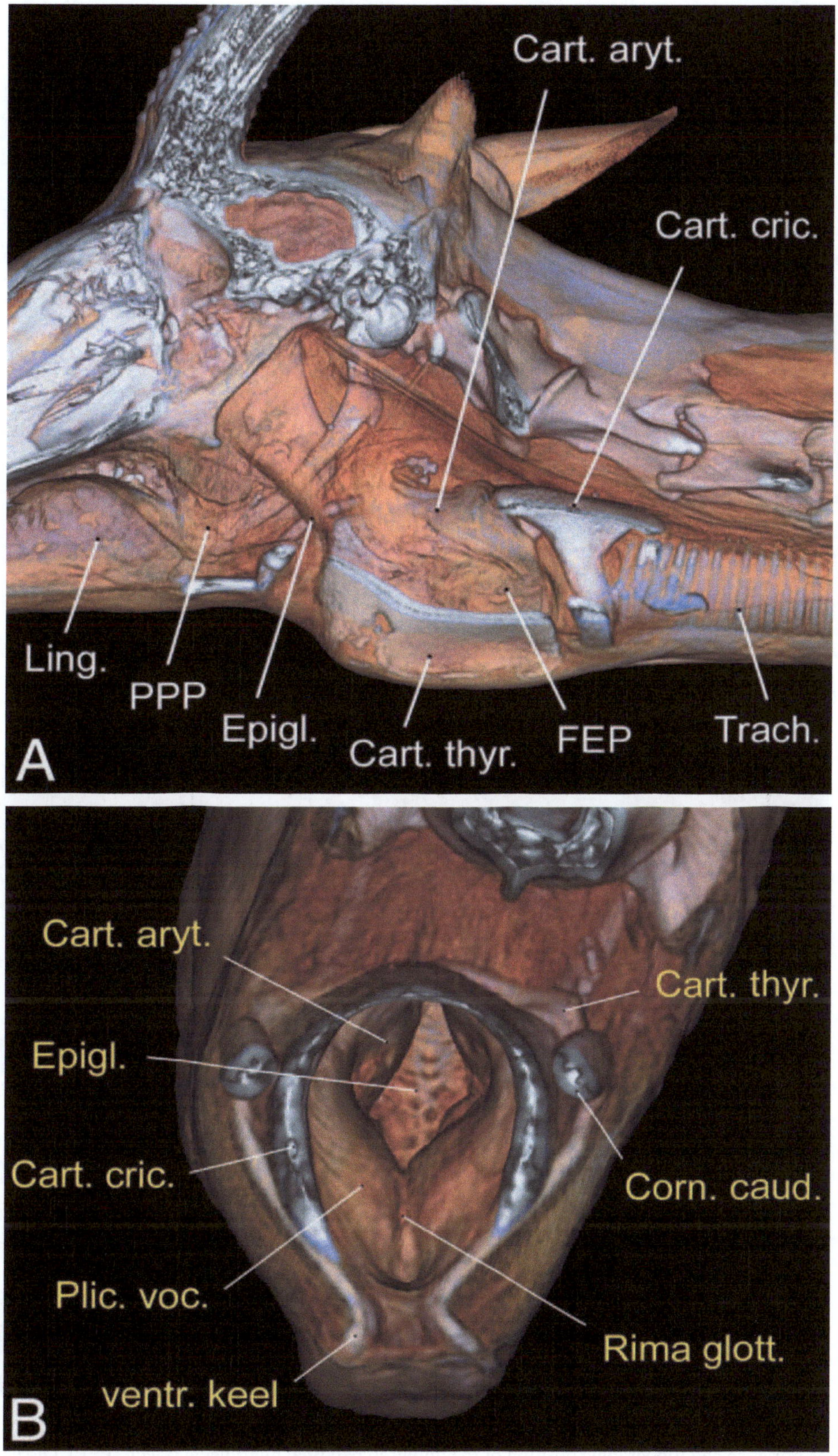

FIG. 1.5. Virtual parasagittal section of CT-based virtual 3D reconstruction of pharyngolaryngeal region of a male Mongolian gazelle (**A**) disclosing the connection of the arytenoid cartilage to the cymbal-shaped fibroelastic pad (FEP) supporting the vocal fold. Virtual transverse section through the larynx of A at the level of the vocal folds, caudal view (**B**). Mediosagittal section of larynx in medial view exposing the right vocal fold (**C**). (**C** alterated after Frey and Riede 2003)

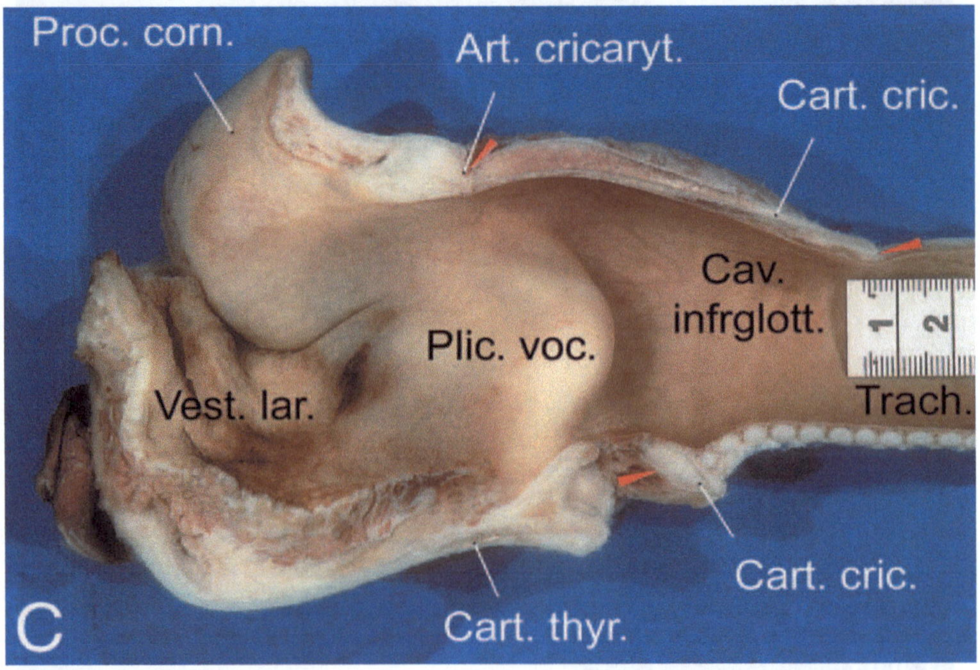

FIG. 1.5. *Continued*

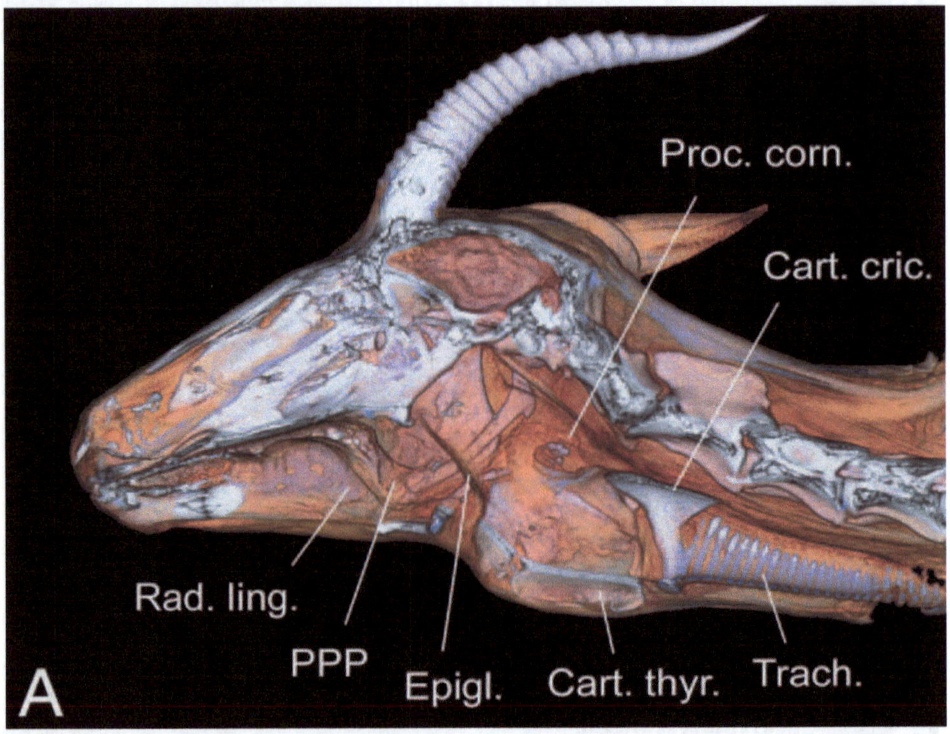

FIG. 1.6. Virtual parasagittal section of CT-based 3D reconstruction of the head of a male Mongolian gazelle (**A**) demonstrating presence and topographic position of the unpaired palatinal pharyngeal pouch (PPP). Dissection confirms its position between the root of the tongue and the epiglottis. In addition, the strongly wrinkled wall of the PPP is revealed (**B**) implying that some sort of inflation might occur. (**B** alterated after Frey and Gebler 2003)

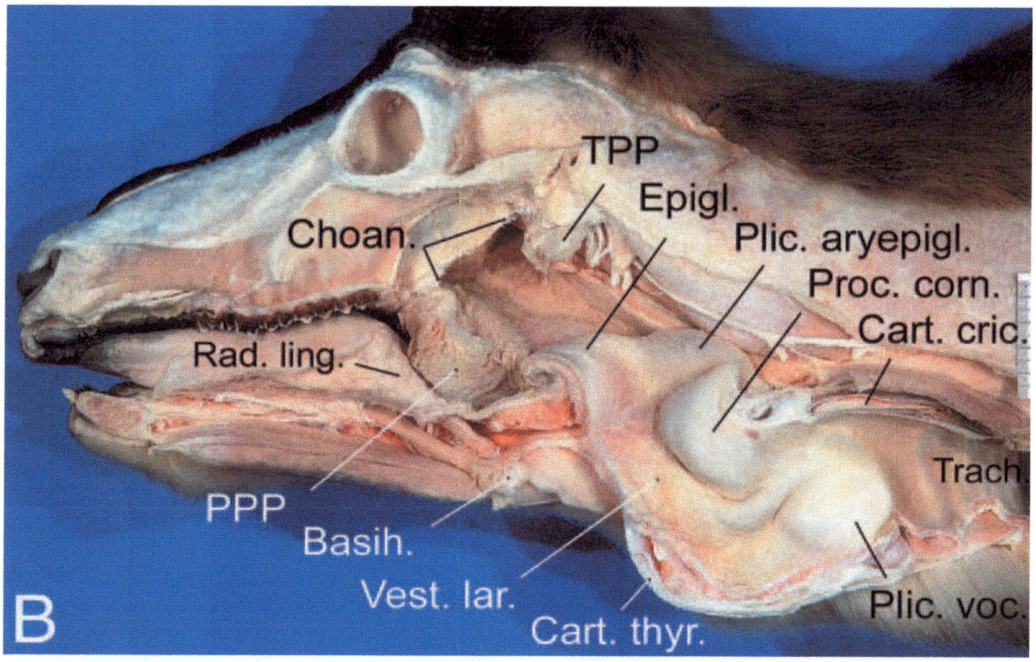

FIG. 1.6. *Continued*

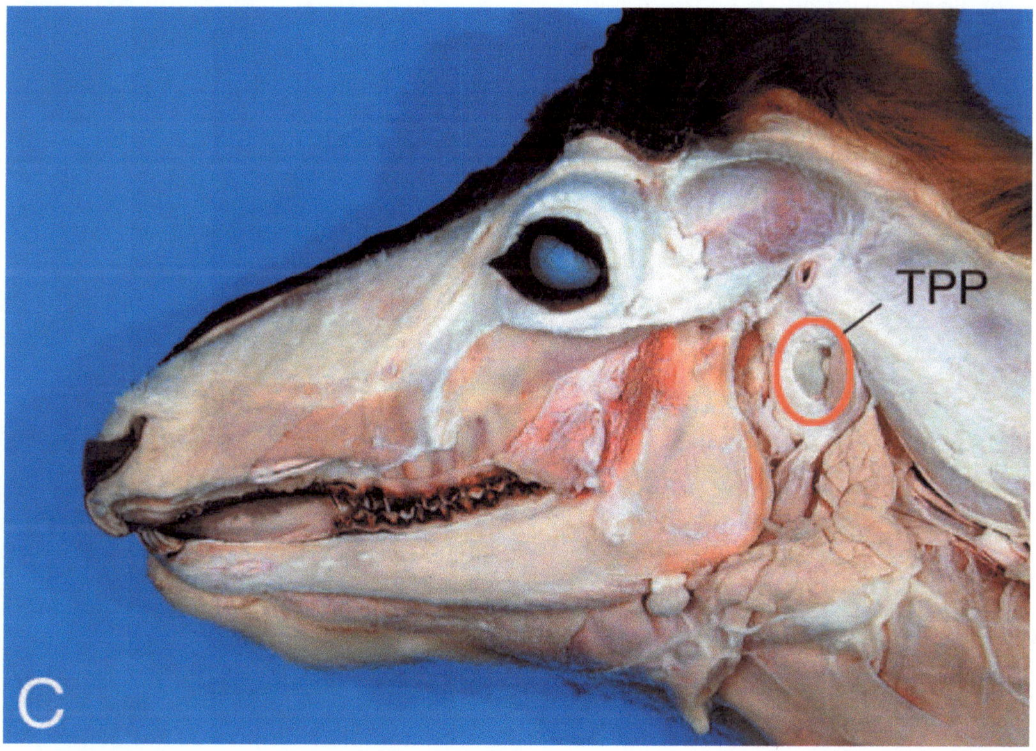

FIG. 1.7. Dissected topographic position of TPP (**C**) between tympanohyoid (rostrally), stylohyoid (laterally) and origin of occipitohyoid muscle (caudally).

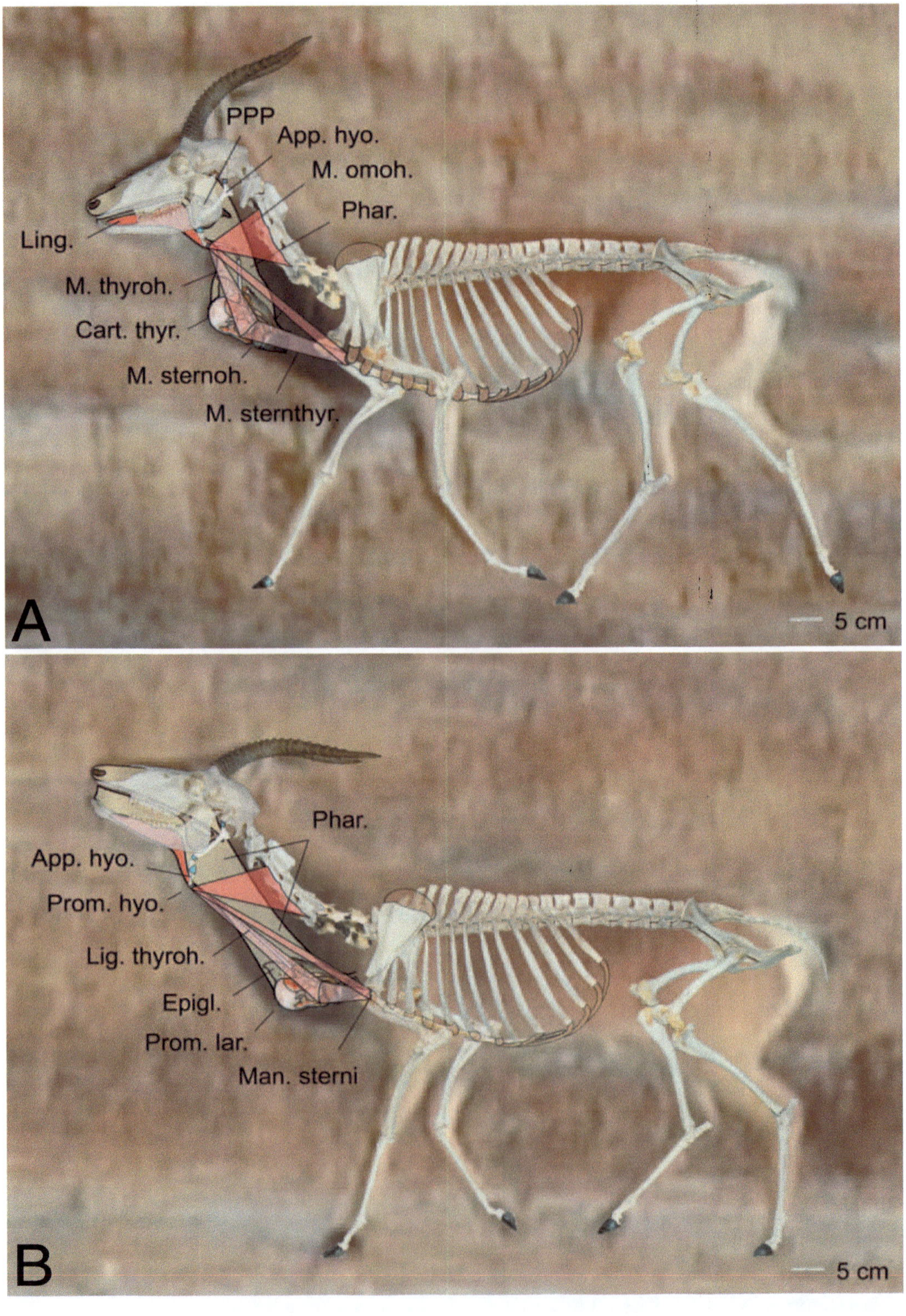

FIG. 1.8. Low resting position of the larynx during the rut (**A**) and maximally retracted larynx during a head-up bark of a male Mongolian gazelle (**B**). In this stage the hyoid apparatus is tilted caudally and ligament-like thyrohyoid connection and pharynx are maximally extended. Skeleton and respective organs were fitted to video single frames.

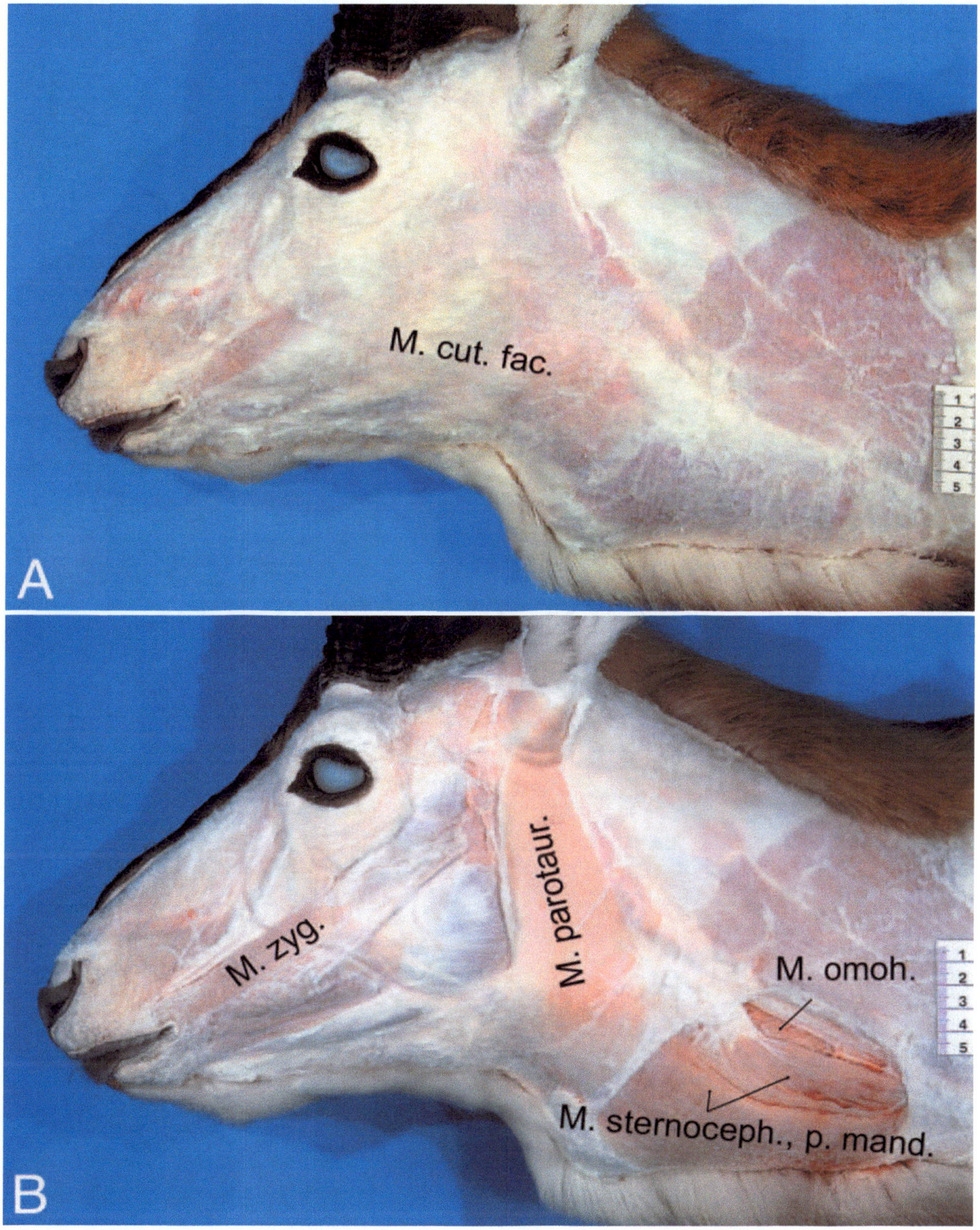

FIG. 1.9. Delicate cutaneous muscle of the head of a male Mongolian gazelle (**A**). Parotidoauricularis muscle, omohyoid muscle and the two portions of the strongly developed sternocephalic muscle after removal of the cutaneous muscle of the head (**B**). The parotidoauricularis muscle supports the epiglottal region in a hammock-like sling. Omohyoid and mandibular portion of sternocephalic muscle lend lateral support to the large and heavy larynx.

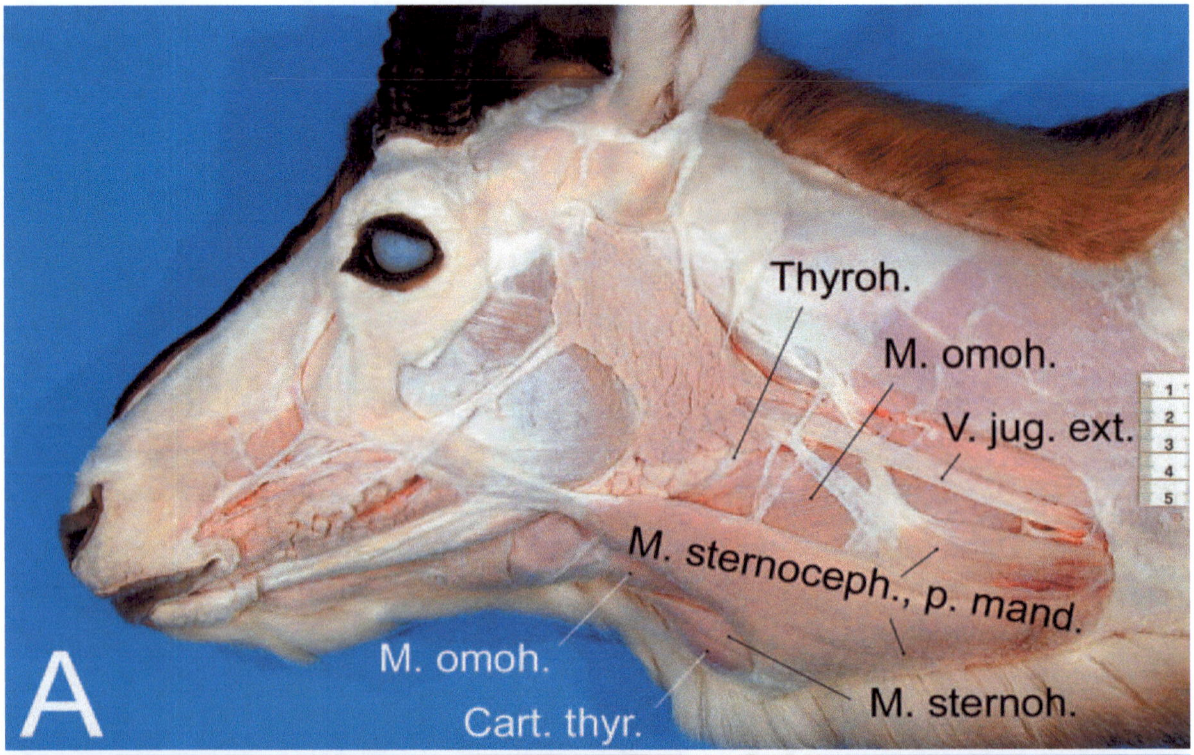

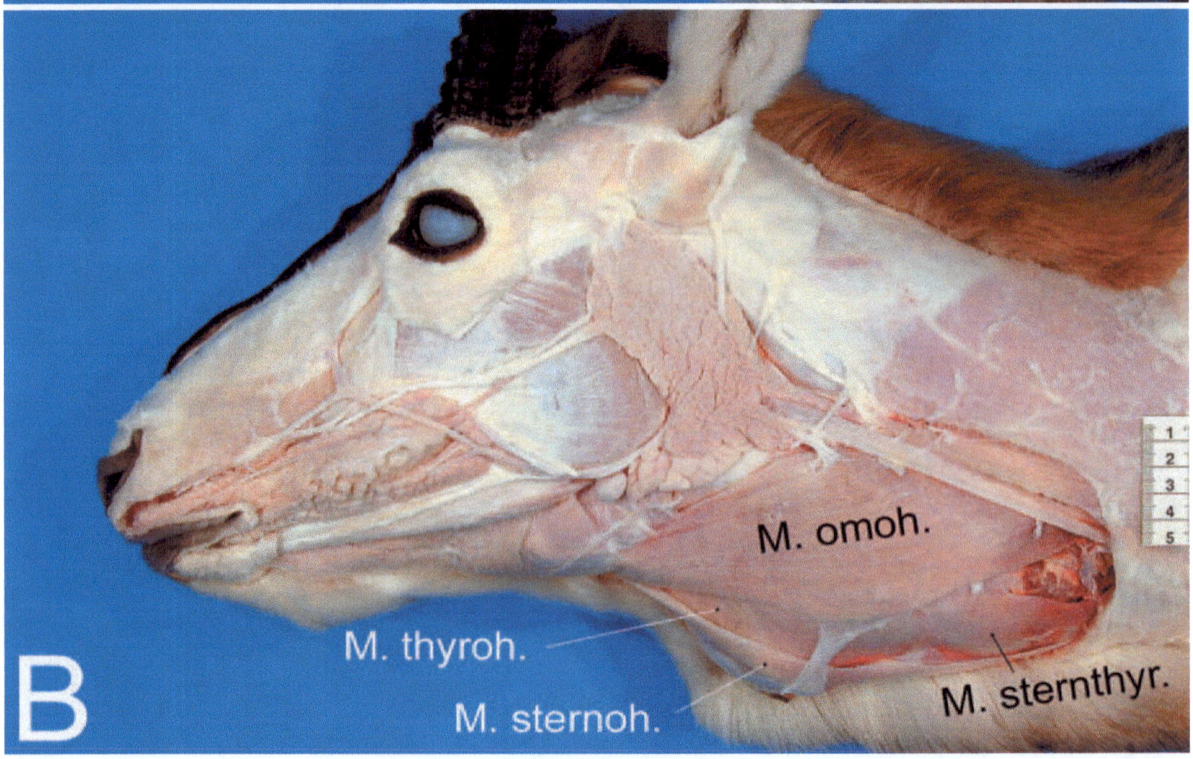

FIG. 1.10. The double-bellied mandibular portion of the sternocephalic muscle has an additional aponeurotic insertion, traversed by the external jugular vein, to the parotid fascia and to the ventral edge of the neck muscles (**A**). After removal of the mandibular portion of the sternocephalic muscle, the broad omohyoid muscle is exposed (**B**). It covers the pharyngolaryngeal region laterally. Ventrally, the omohyoid muscle is connected with its contralateral counterpart by connective tissue.

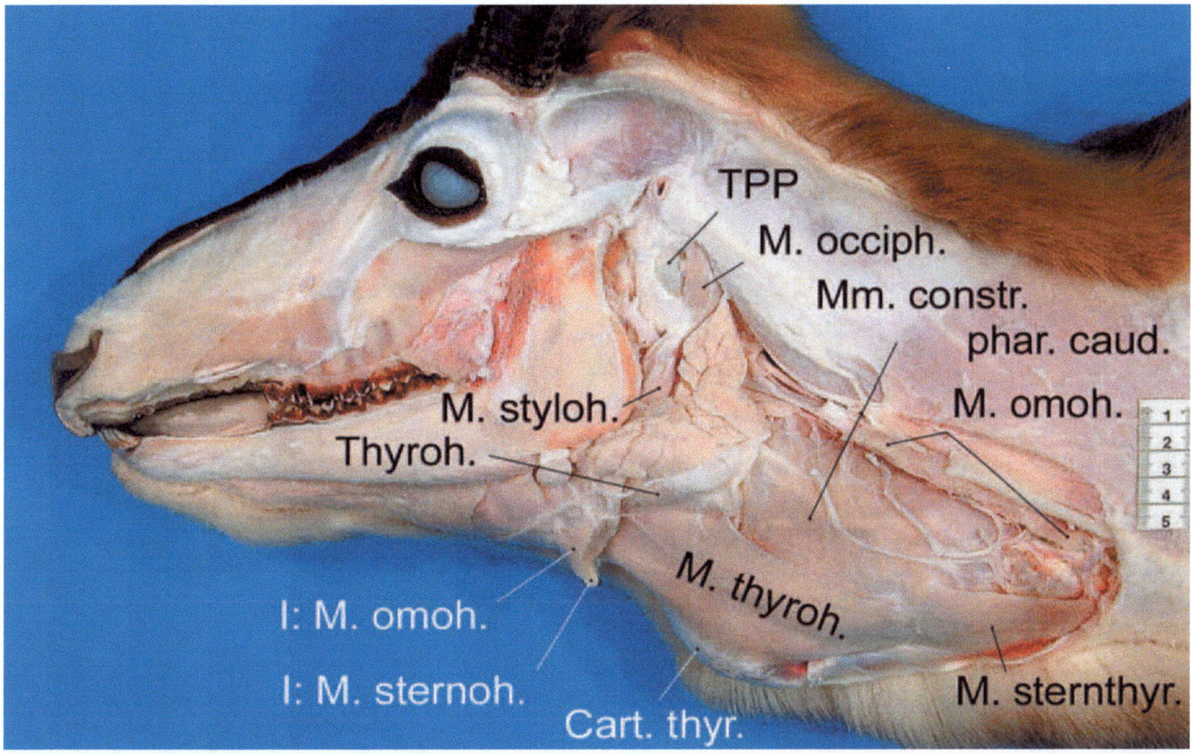

FIG. 1.11. Deep musculature of the laryngeal region of a male Mongolian gazelle after removal of the omohyoid muscle. The larynx is connected to the sternum by the sternothyroid muscle and to the hyoid apparatus by the thyrohyoid muscle.

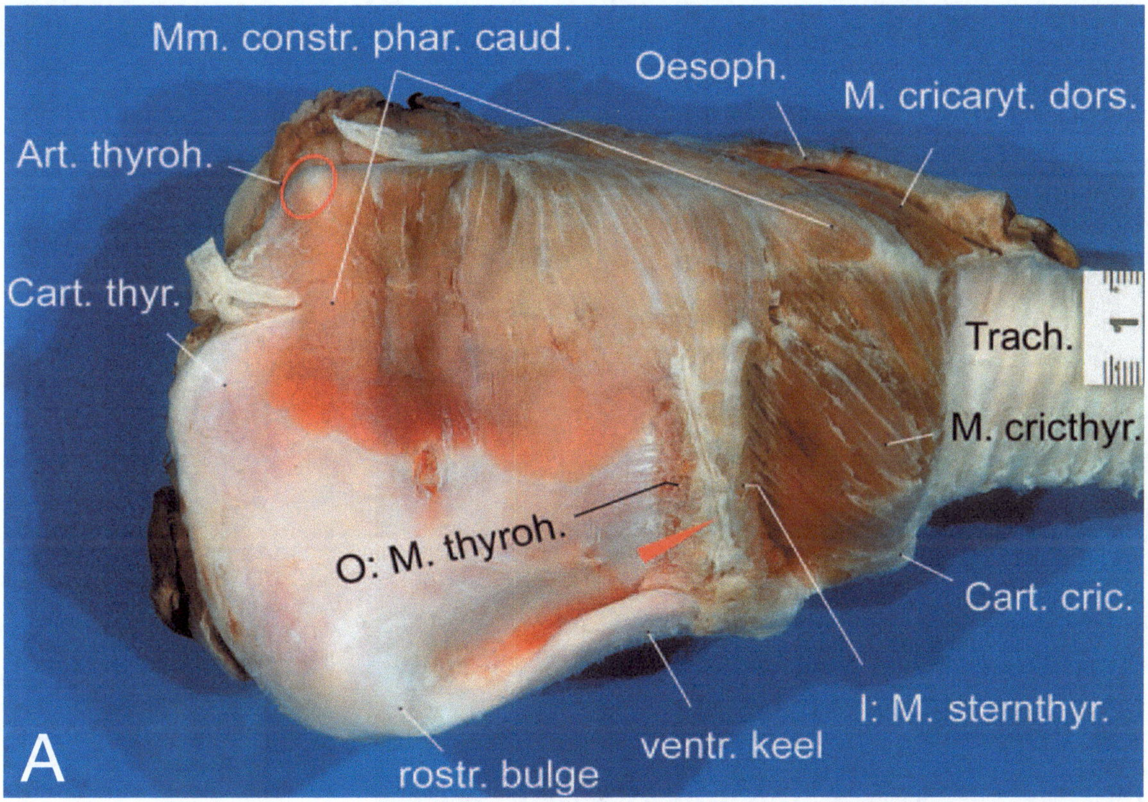

FIG. 1.12. The intrinsic laryngeal muscles of a male Mongolian gazelle: the most superficial one is the cricothyroid muscle (**A**). After removal of the left half of the thyroid cartilage, the transverse arytenoid muscle, the dorsal and lateral cricoarytenoid muscles and the thyroarytenoid muscle are exposed. The latter is divided into a rostral venticularis and a caudal vocalis muscle. Between them the two-chambered inflatable LLV projects laterally (**B**). (**B** alterated after Frey and Gebler 2003 and Frey and Riede 2003)

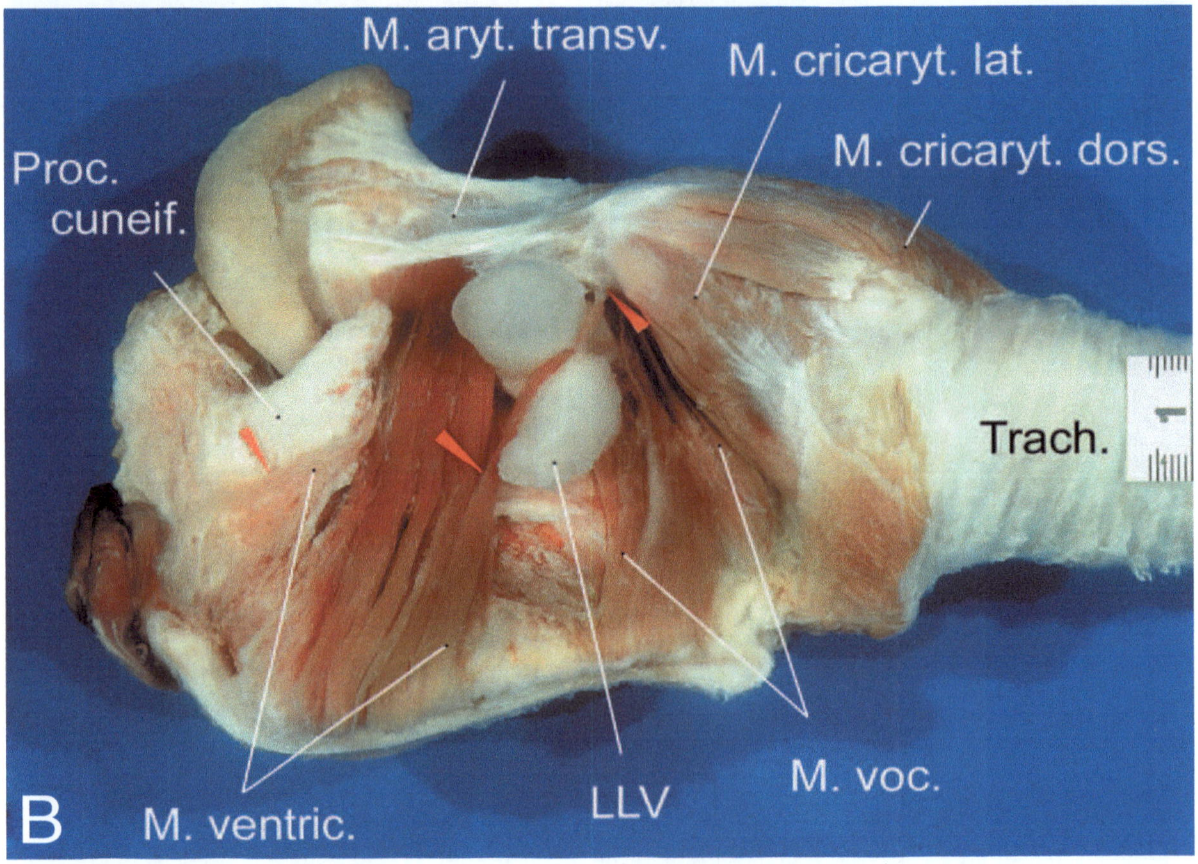

FIG. 1.12. *Continued*

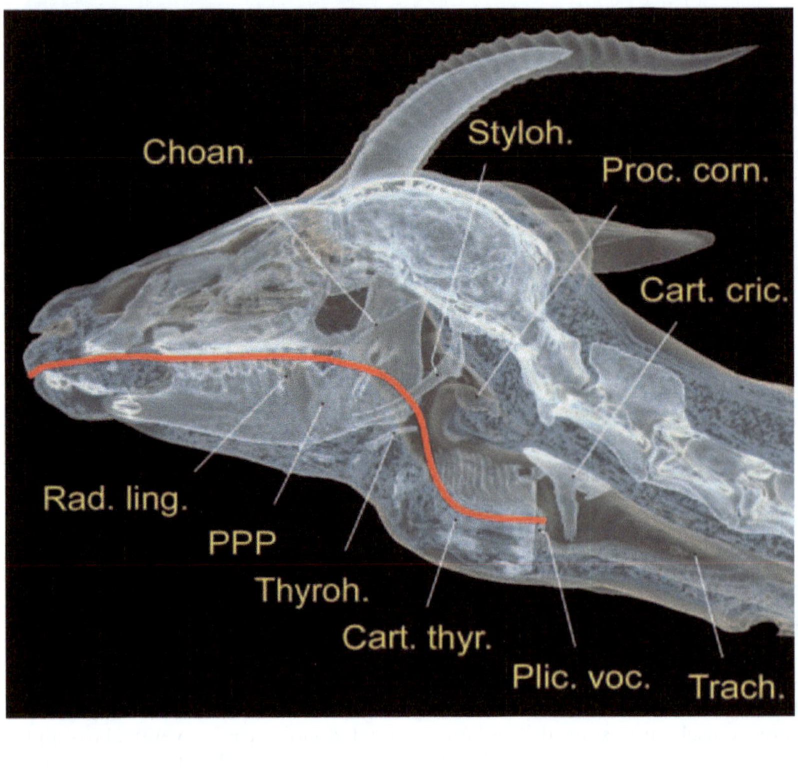

FIG. 1.13. Measurement of vocal tract length (VTL = lips to glottis *in red*) in a virtual mediosagittal section of a computed 3D reconstruction of the head of a male Mongolian gazelle. Owing to the formation of a fibroelastic pad, the vocal fold can be clearly identified between the caudal edge of the thyroid cartilage and the rostral edge of the cricoid cartilage. (alterated after Frey and Gebler 2003)

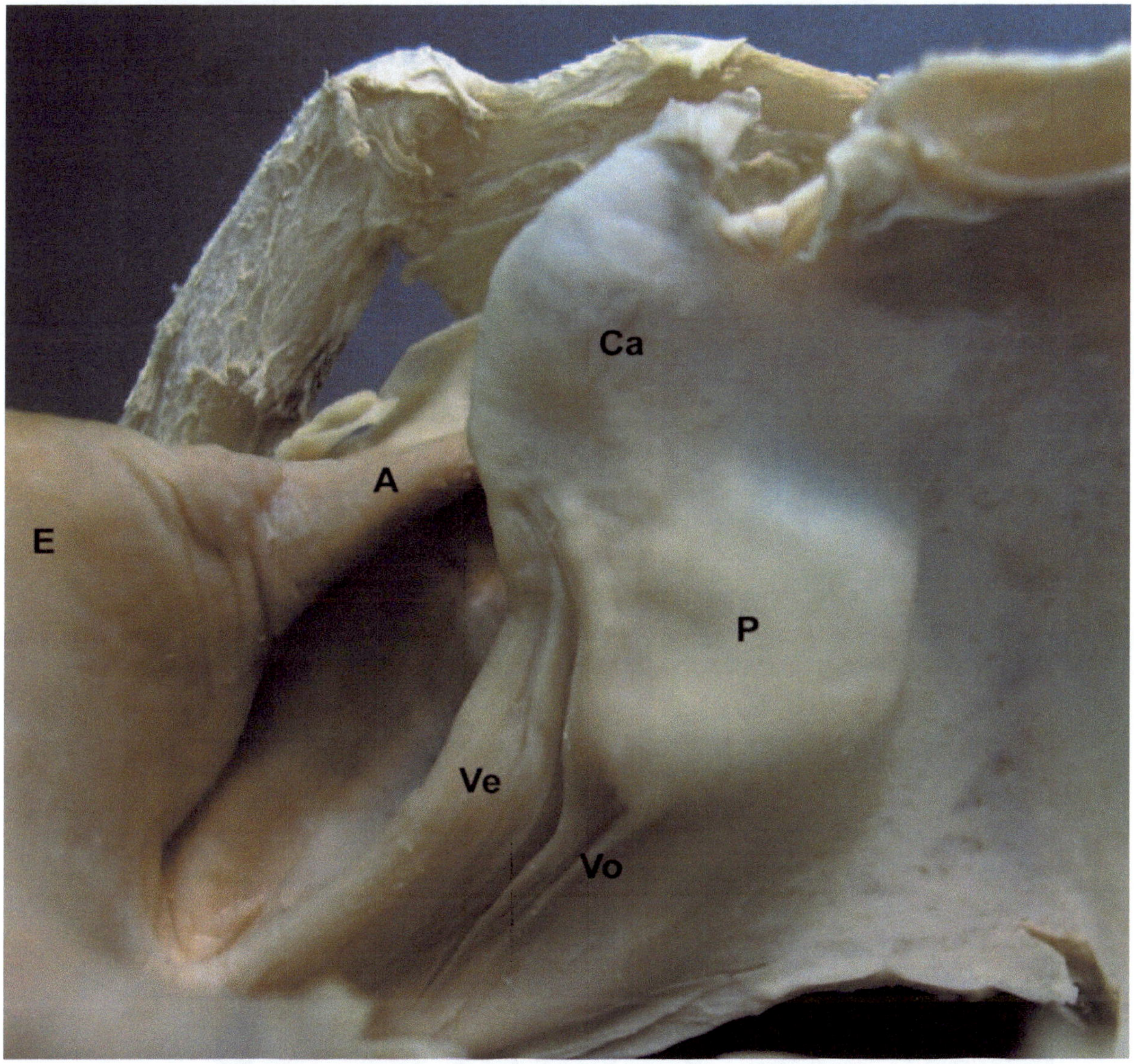

FIG. 2.1. Larynx of the tiger. Medial view. *E*, epiglottis; *A*, Plica aryepiglottica; *Ca*, Cartilago arytenoidea (covered by mucosa); *P*, pad; *Ve*, Plica vestibularis; *Vo*, Plica vocalis.

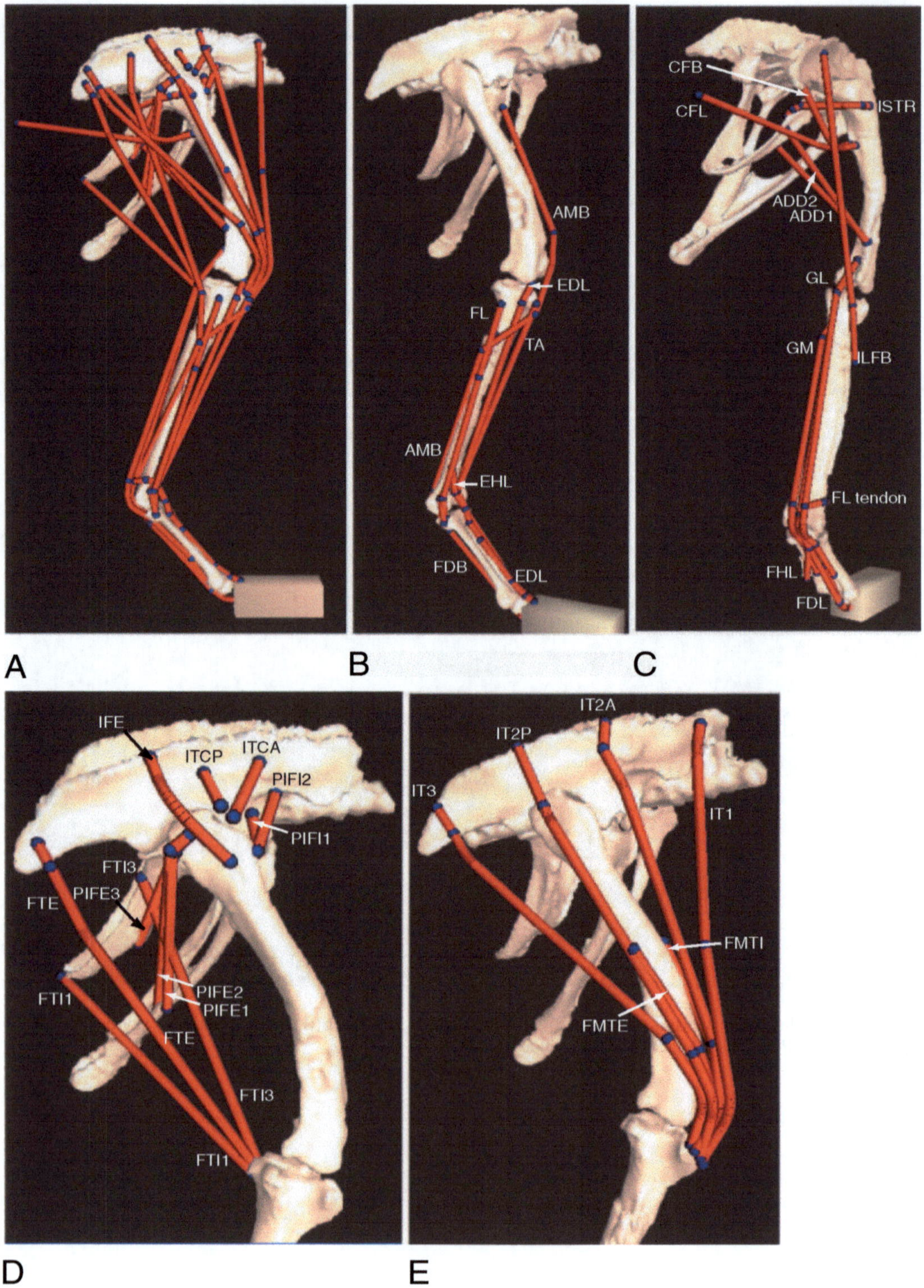

FIG. 3.2. Muscles (*red lines*) of *Velociraptor* right hind limb model, in lateral view with all shown (**A**), oblique craniolateral view of lower limb muscles (**B**), oblique caudolateral view of some hip extensor and lower limb muscles (**C**), dorsolateral view closeup of hamstring (flexor cruris complex), external puboischiofemoral, and deep dorsal thigh muscles (**D**), and craniolateral view closeup of M. iliotibialis and M. femorotibialis knee extensor muscles (**E**). See Table 3.2 for muscle abbreviations.

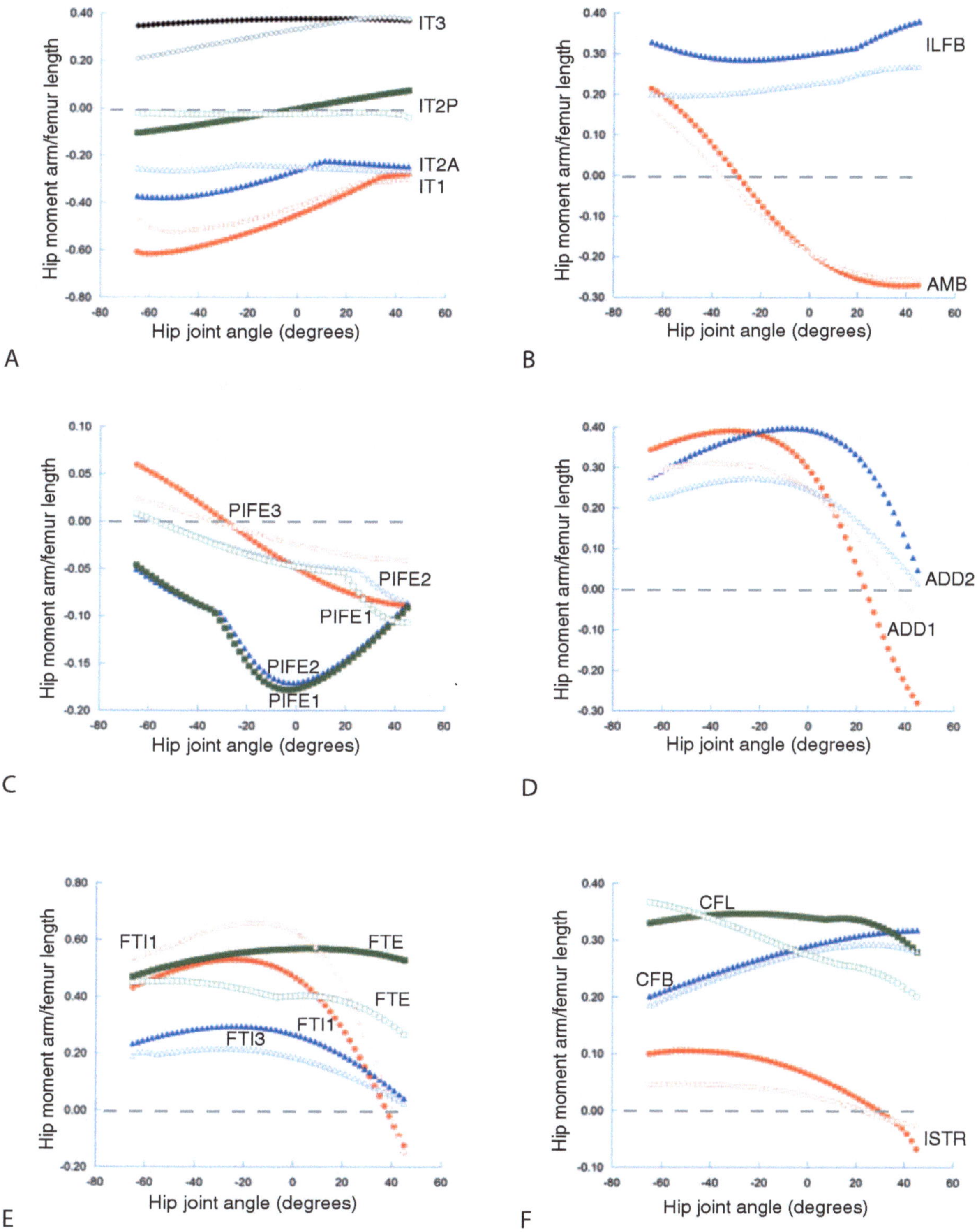

Fig. 3.3. Hip flexor/extensor moment arm-hip joint angle relationships for major hip muscles in *Velociraptor* (*open symbols*) and *Tyrannosaurus* (*filled symbols*). Moment arms are normalized by femur length (see text and Table 3.1) for size-independent comparisons. Positive moment arms are extensor; negative are flexor. A hip joint angle of 0° represents a vertical femur; negative values are extension past that point and positive values are flexion (femoral protraction). **A**, M. iliotibialis 1 – 3 (IT1 – 3). **B**, M. iliofibularis (ILFB) and M. ambiens (AMB). **C**, Mm. puboischiofemorales externi 1 – 3 (PIFE1 – 3). **D**, Mm. adductores femorii 1 + 2 (ADD1 + 2). **E**, Mm. flexores tibiales interni 1 + 3 (FTI1 + 3) et externus (FTE), and **F**, M. ischiotrochantericus (ISTR) and Mm. caudofemorales brevis et longus (CFB, CFL).

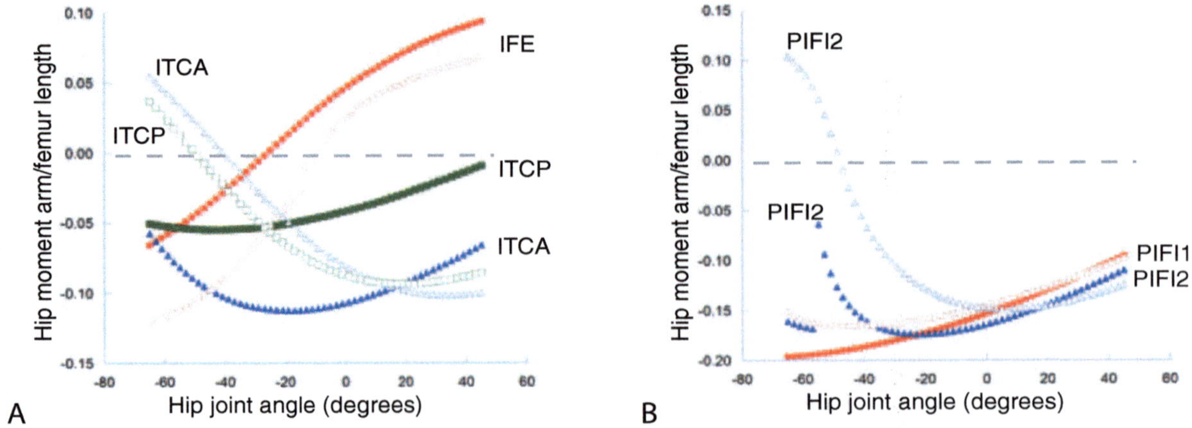

FIG. 3.4. Additional hip moment arm-joint angle relationships as in Fig. 3.3. **A**, M. iliofemoralis externus (IFE) and M. iliotrochantericus caudalis (ITCA, ITCP). **B**, Mm. puboischiofemoralis internus 1 + 2 (PIFI1 + 2).

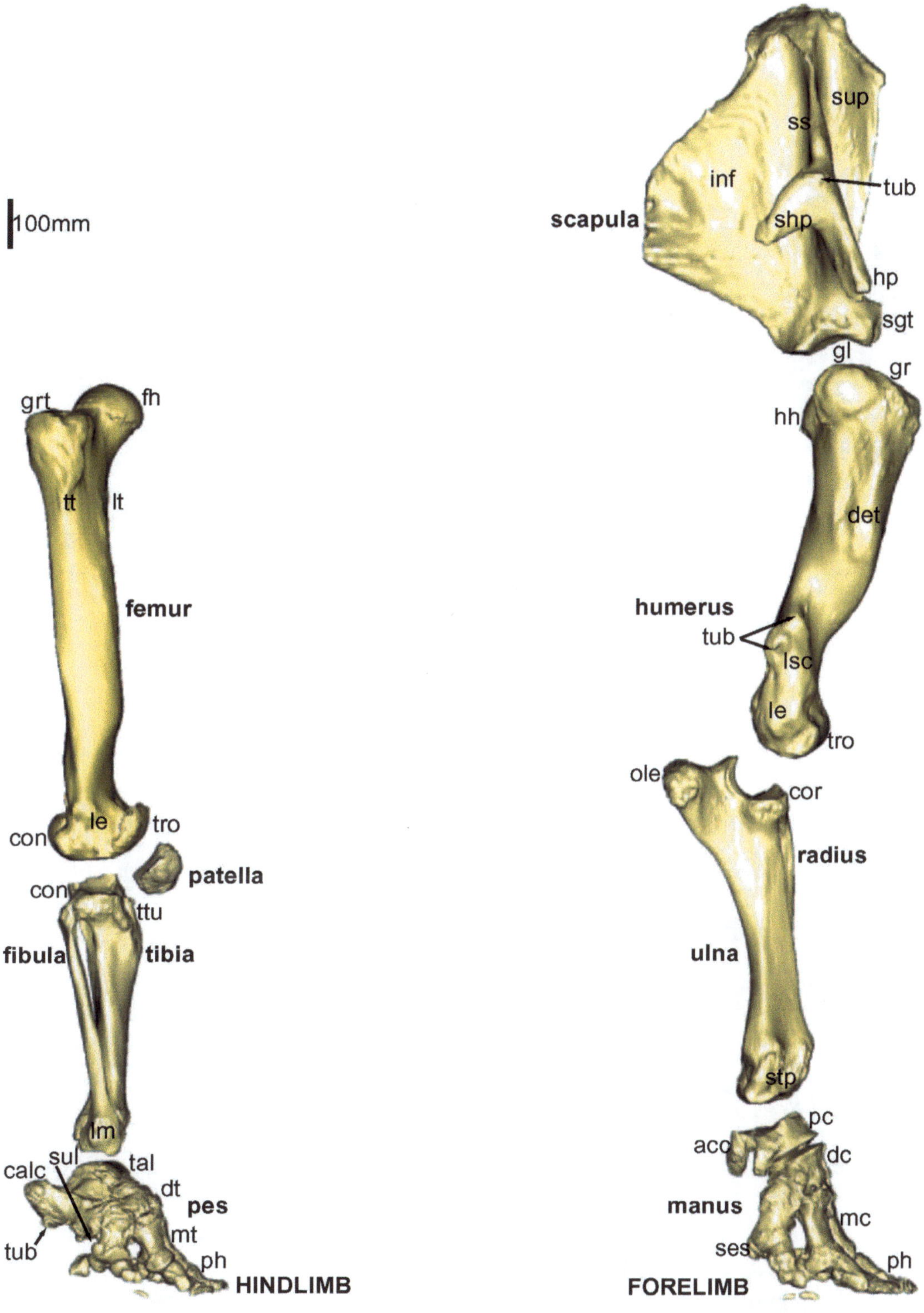

FIG. 3.5. Bones of the right fore and hind limb of our study specimen of *Elephas maximus*, three-dimensionally rendered in MIMICS software from CT scans. In lateral view. The major bones are moved slightly out of articulation so more of their surfaces can be seen. Abbreviations: acc = accessory carpal, calc = calcaneus, con = condyle, cor = medial coronoid process, dc = distal carpals (I-V), det = deltoid tuberosity, dt = distal tarsals (I-V), fh = femoral head, gl = glenoid surface, gr = greater tubercle, grt = greater trochanter, hh = humeral head, hp = hamate process, inf = infraspinous fossa, le = lateral epicondyle, lm = lateral malleolus, lsc = lateral supracondylar crest, lt = lesser trochanter, mc = metacarpals (I-V), mt = metatarsals (I-V), ole = olecranon process, pc = proximal carpals (I-V), ph = phalanges (I-V), ses = sesamoid(s), sgt = supraglenoid tubercle, shp = suprahamate process, ss = scapular spine, stp = styloid process, sul = sulcus, sup = supraspinous fossa, tal = talus, tro = trochlear surface, tt = third trochanter, tub = tubercle, ttu = tibial tuberosity.

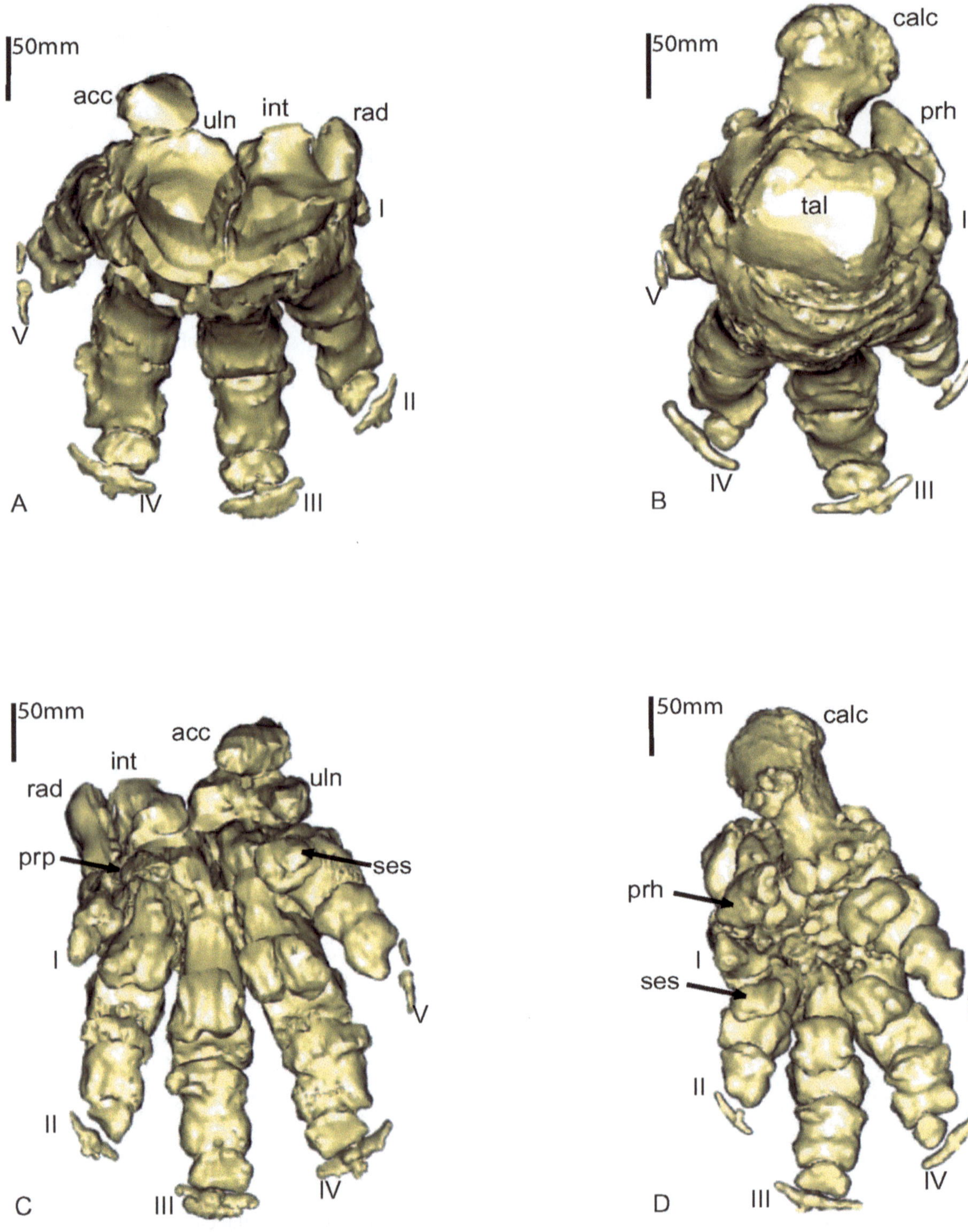

FIG. 3.6. Articulated bones of the right manus (**A**, **C**) and pes (**B**, **D**), in palmar (**A**, **B**) and planter (**C**, **D**) views. Abbreviations as in Fig. 3.5 caption plus: I-V = digits I-V, int = intermediate carpal, prh = prehallux, prp = prepollex, rad = radial carpal, tal = talus, uln = ulnar carpal.

FIG. 3.7. 'Predigits' of the right manus and pes: prehallux in medial (**A**) and lateral (**B**) views, prepollex in cranial (**C**) and caudal (**D**) views, and dorsomedial views (not to scale) of manus (**E**) and pes (**F**), with articulations shown. Abbreviations as in Figs 3.5,3.6 captions plus: cen = central tarsal, MC1 = metacarpal I, MT1 = metatarsal I, prh-d = distal segment of prehallux, prh-p = proximal segment of prehallux, T1 = (distal) tarsal I.

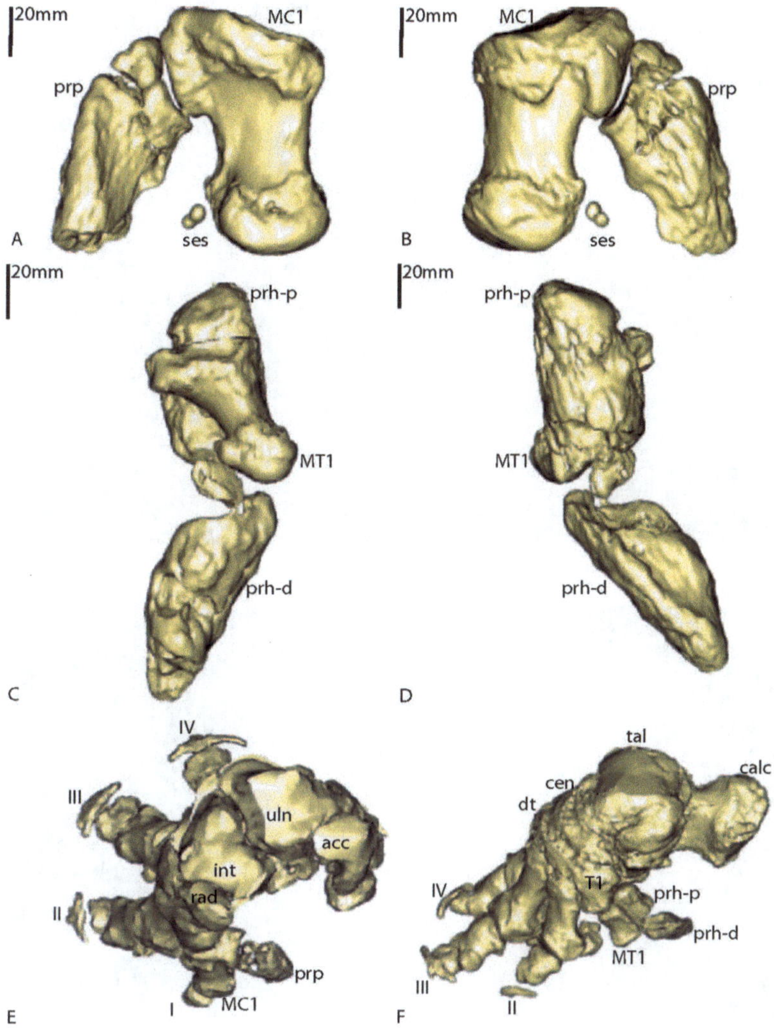

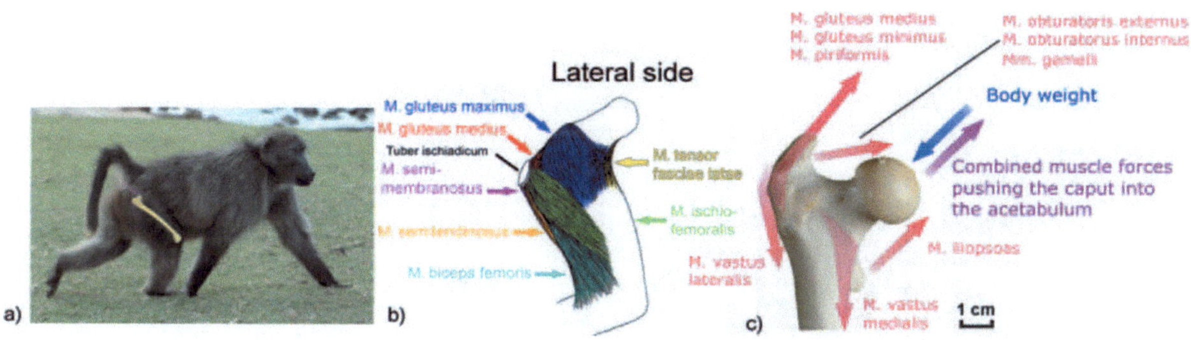

FIG. 4.2. (**a**) Walking chacma baboon (*Papio ursinus*) with outlined femur; photograph by Larissa Swedell. (**b**) Position of lateral thigh muscles which load the femur; changed after Swindler and Wood (1973). (**c**) Directions of action of proximal thigh muscles; after Scherf (2007), photograph by Anika Hebs, Senckenberg Research Institute (Germany).

FIG. 4.3. High resolution CT System Rayscan 200 at the Foundry Technology centre of the Aalen University of Applied Sciences; Scherf (2007).

FIG. 4.4. Leaping Hanuman langur (*Semnopithecus entellus*); photograph by Paul Winkler, Georg-August-Universität Göttingen (Germany).

FIG. 4.5. Running adult female chacma baboon (*Papio ursinus*) with infant on belly; photograph by Larissa Swedell, City University of New York (USA).

Fig. 4.6. Slowly brachiating gibbon (*Hylobates lar*) in Khao Yai National Park, Thailand; photograph by Claudia Barelli, Max Planck Institute for Evolutionary Anthropology (Germany).

Fig. 4.7. Walking humans; © Max Planck Institute for Evolutionary Anthropology (Germany).

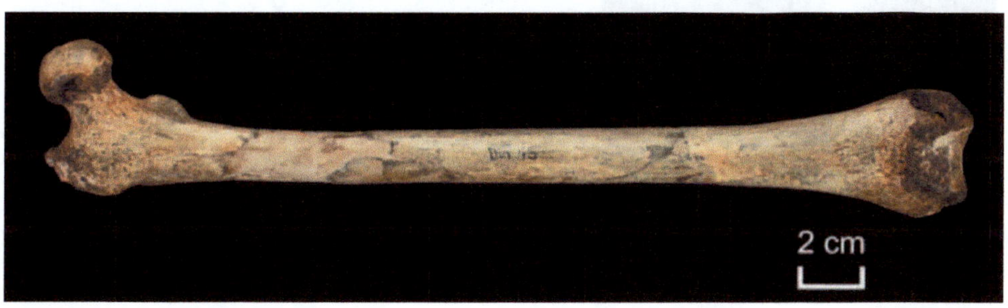

Fig. 4.8. Ventral view of the right femur of *Paidopithex rhenanus* (Din45); after Scherf (2007), photograph by Anika Hebs, Senckenberg Research Institute (Germany).

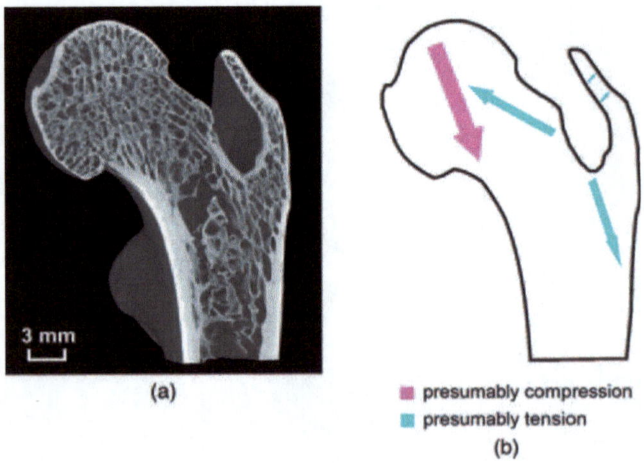

FIG. 4.10. (**a**) Ventral view of the trabecular architecture of the left femur of *Alouatta seniculus* (Senckenberg Research Institute, 25 544 ♀), Scherf (2007), (**b**) Outline of trabecular alignment.

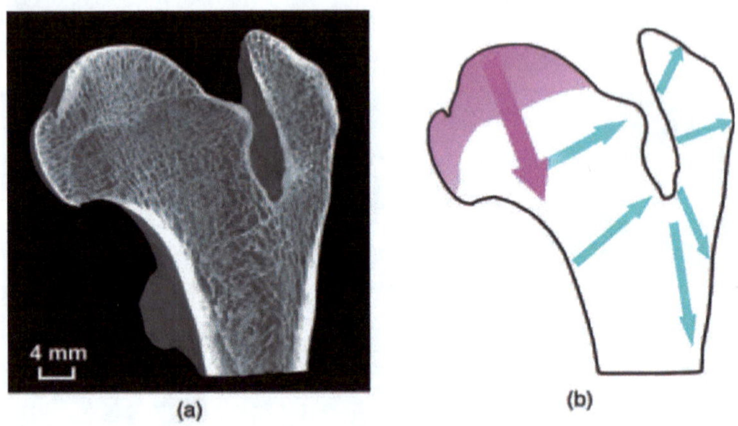

FIG. 4.11. (**a**) Ventral view of the trabecular architecture of the left femur of *Semnopithecus entellus* (Institute of Zoology of the University of Hamburg, 4734, sex was not registered) most mature specimen, Scherf (2007), (**b**) Outline of trabecular alignment.

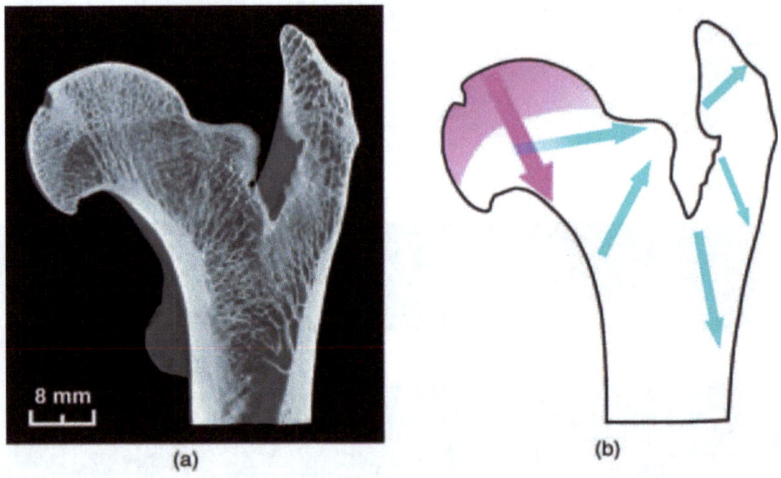

FIG. 4.12. (**a**) Ventral view of the trabecular architecture of the left femur of *Papio hamadryas* (Dr. Senckenbergische Anatomie, Ha VIII 3 ♂), Scherf (2007), (**b**) Outline of trabecular alignment.

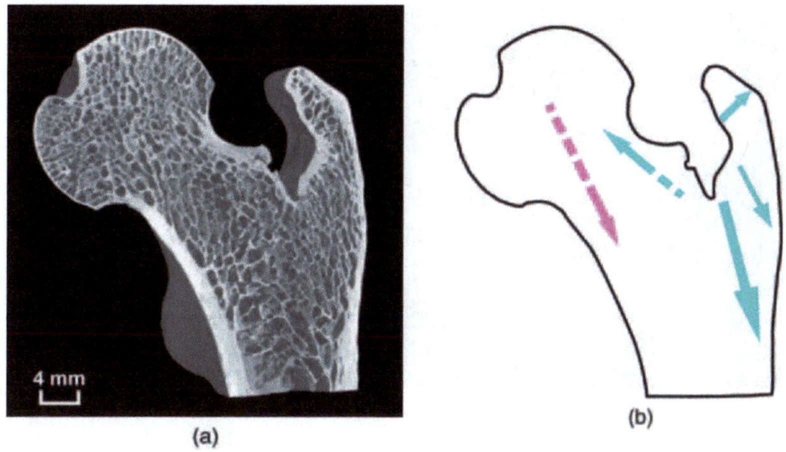

FIG. 4.13. (**a**) Ventral view of the trabecular architecture of the left femur of *Hylobates syndactylus* (Dr. Senckenbergische Anatomie, 52.36. ♀), Scherf (2007), (**b**) Outline of trabecular alignment.

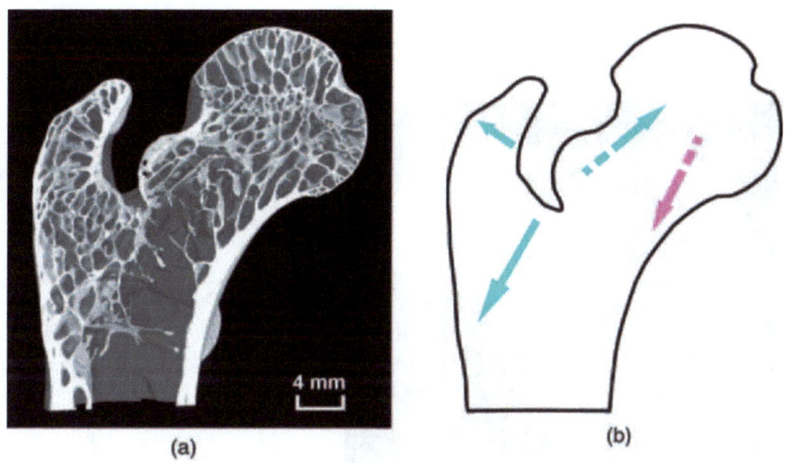

FIG. 4.14. (**a**) Ventral view of the trabecular architecture of the right femur of *Hylobates lar moloch* (Senckenberg Research Institute, 47 979 ♂), Scherf (2007), (**b**) Outline of trabecular alignment.

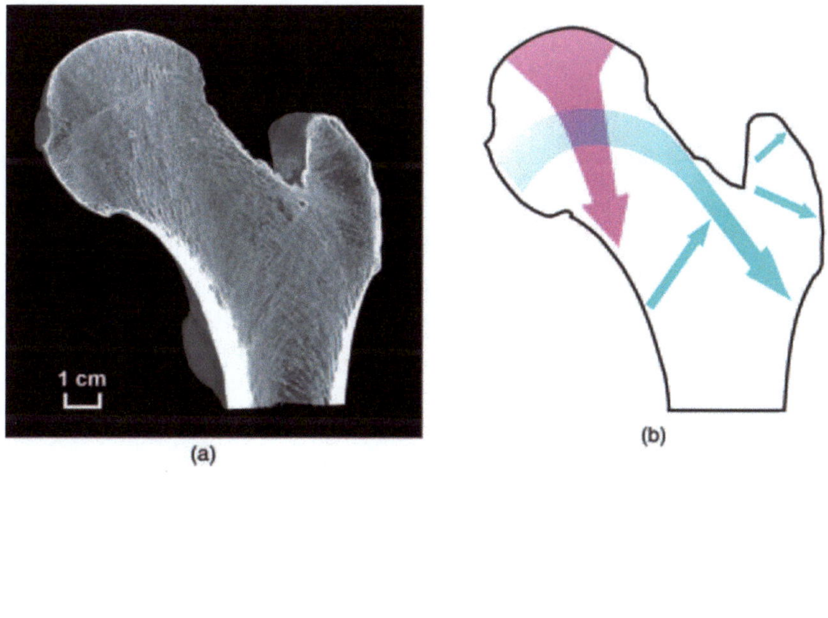

FIG. 4.15. (**a**) Ventral view of the trabecular architecture of the left femur of *Homo sapiens* (Senckenberg Research Institute, 21 ♂), Scherf (2007), (**b**) Outline of trabecular alignment.

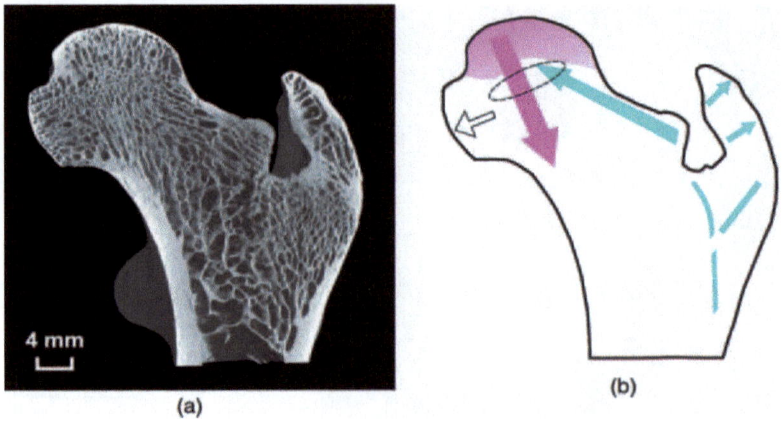

FIG. 4.16. (**a**) Ventral view of the trabecular architecture of the left femur of *Pliopithecus vindobonensis* (Naturhistorisches Museum Basel, O.E. 559), Scherf (2007), (**b**) Outline of trabecular alignment.

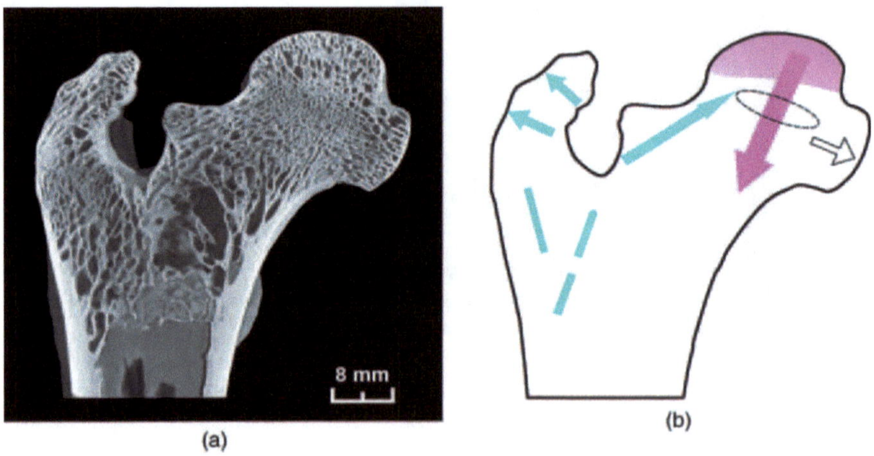

FIG. 4.17. (**a**) Ventral view of the trabecular architecture of the right femur of *Paidopithex rhenanus* (Hessisches Landesuseum Darmstadt, Din 45), Scherf (2007), (**b**) Outline of trabecular alignment.

TABLE 4.3. Outline of trabecular arrangement according to different gross loading types.

High hind limb loading	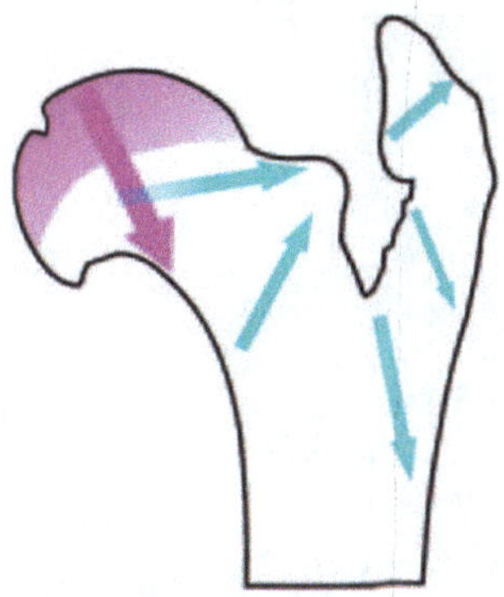 *Papio hamadryas*	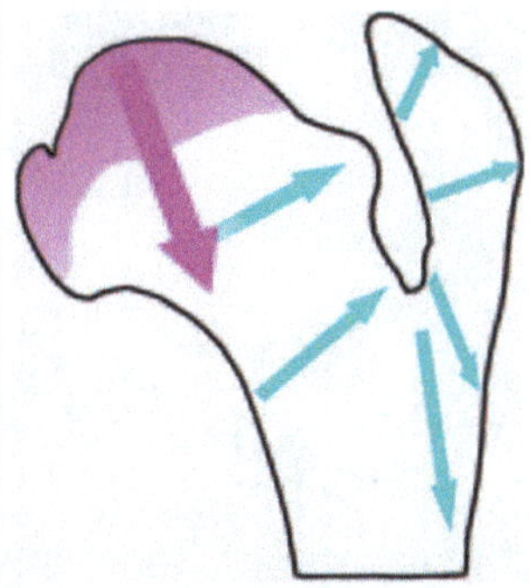 *Semnopithecus entellus*	
Medium hind limb loading	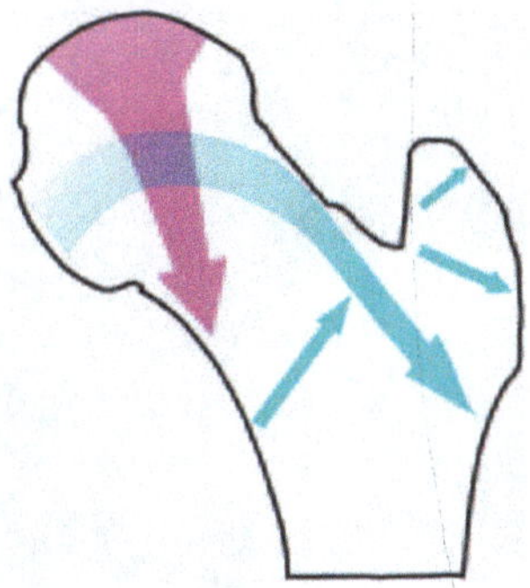 *Homo sapiens*		
Low hind limb loading	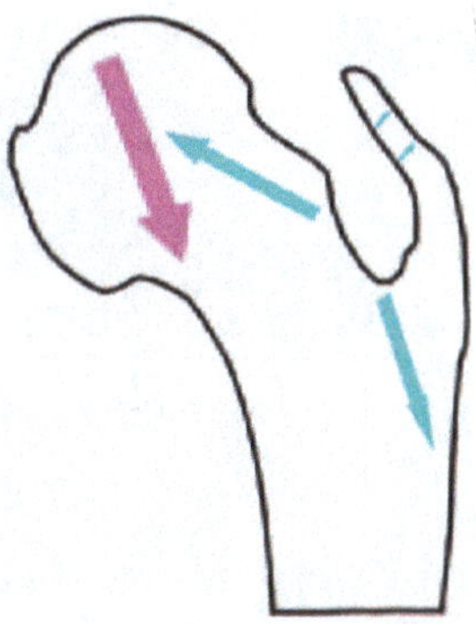 *Alouatta seniculus*	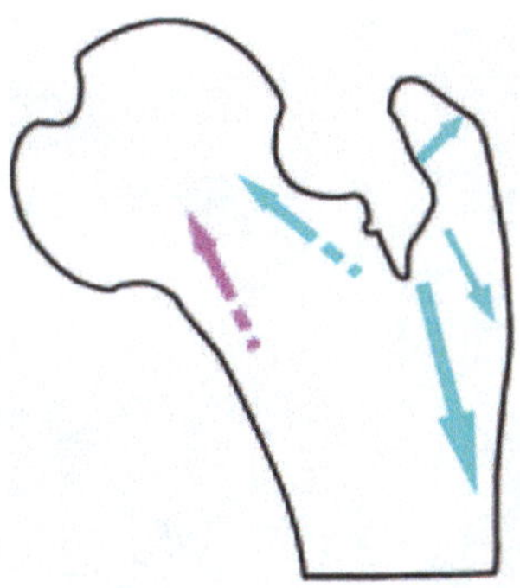 *Hylobates syndactylus*	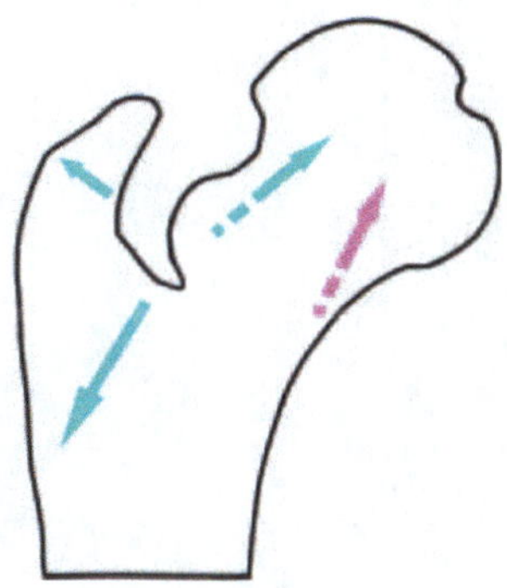 *Hylobates lar moloch*
Fossil species	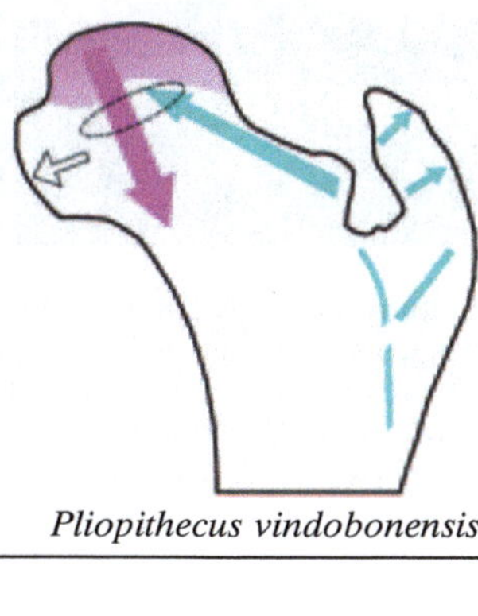 *Pliopithecus vindobonensis*	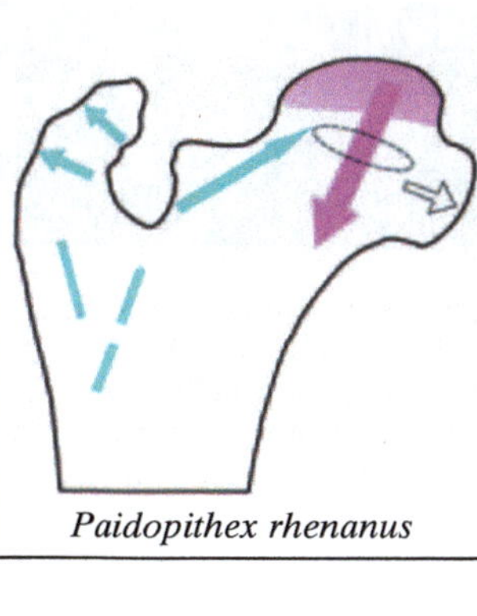 *Paidopithex rhenanus*	

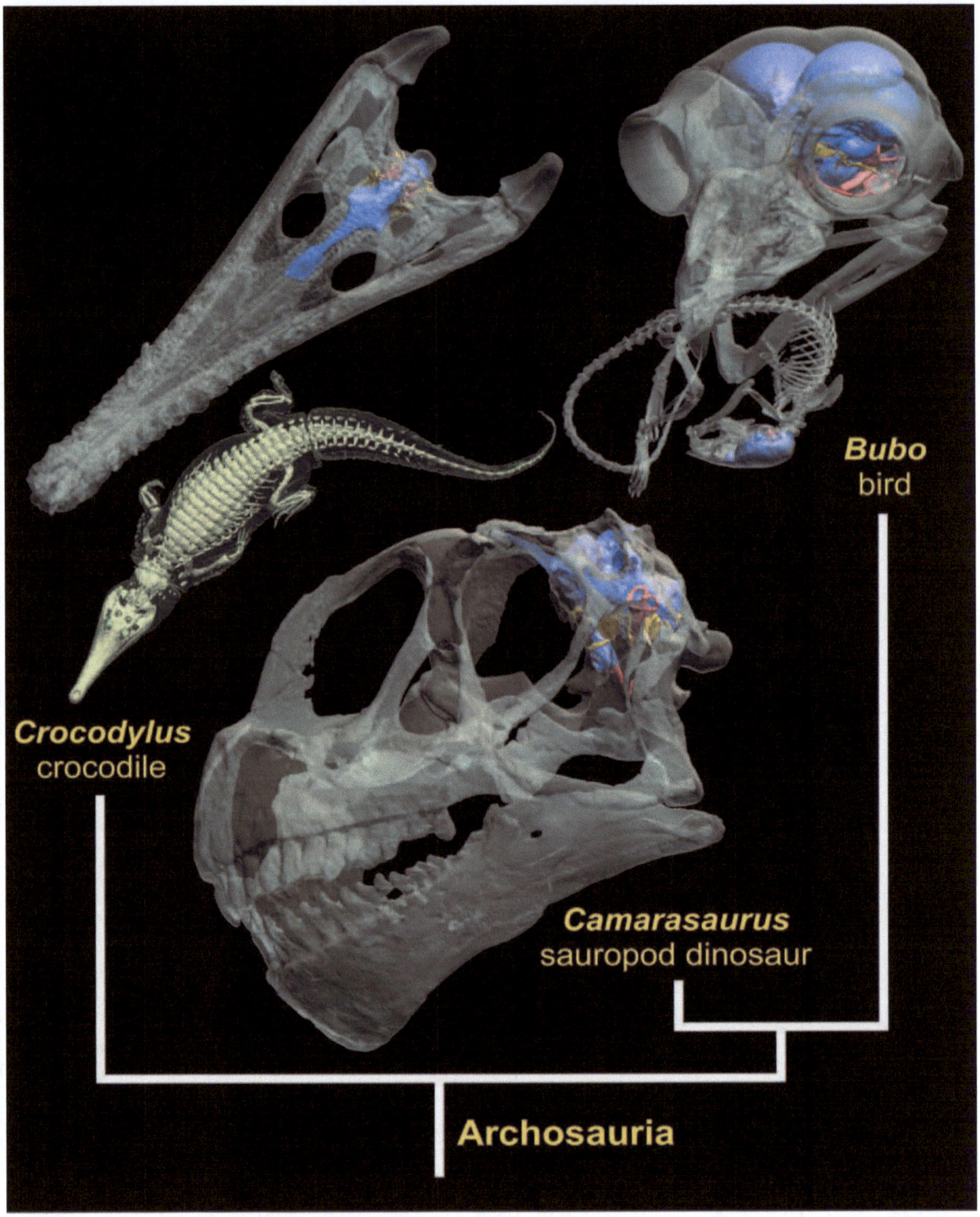

FIG. 6.1. The extant phylogenetic bracket (Witmer 1995) of nonavian dinosaurs, as exemplified by the sauropod dinosaur *Camarasaurus lentus* (CM 11338, in left lateral view). The extant outgroups of the fossil taxa provide information on attributes such as soft-tissue morphology and behavior that are not preserved in the fossils. The extant outgroups of nonavian dinosaurs are crocodilians (as exemplified by *Crocodylus johnstoni*, OUVC 10425, in dorsal view) and birds (as exemplified by *Bubo virginianus*, OUVC 10220, in left rostrolateral view with *Mus musculus* [OUVC 10449] in its jaws). The images are surfaces renderings of CT scan data, with the skull rendered semitransparent revealing the cranial endocast (*in blue*) and other structures.

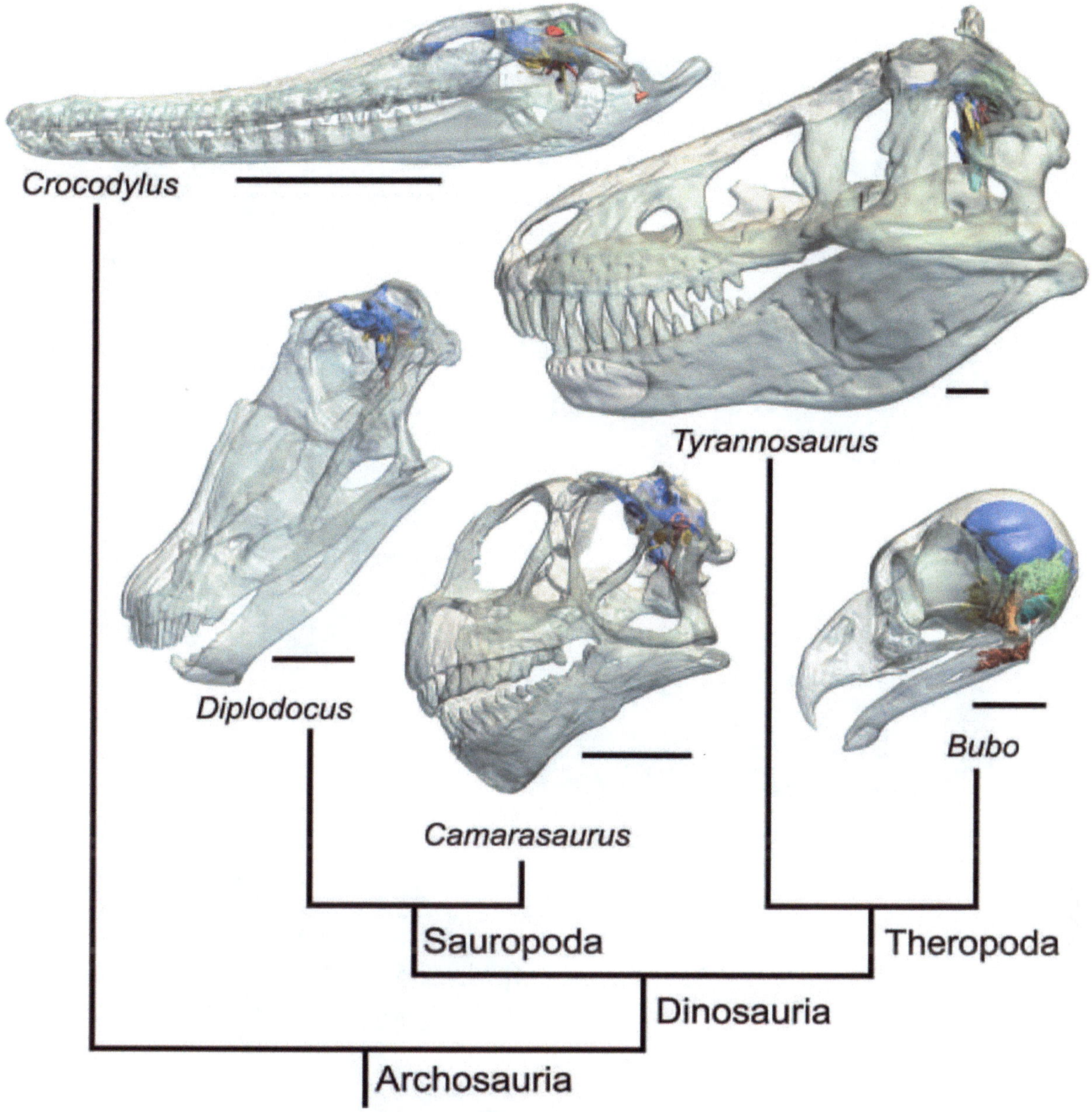

FIG. 6.2. Diagram of the phylogenetic relationships of the principal taxa discussed in the text. From left, the extant crocodilian *Crocodylus johnstoni* (OUVC 10425), the diplodocoid sauropod *Diplodocus longus* (CM 11161), the macronarian sauropod *Camarasaurus lentus* (CM 11338), the basal coelurosaurian theropod *Tyrannosaurus rex* (AMNH 5117), and the avian theropod (bird) *Bubo virginianus* (OUVC 10220). These sauropods and theropods are saurischian dinosaurs. The images are surface renderings of CT scan data. Bony skulls are depicted as transparent, revealing such internal structures as the brain cast, labyrinth, and pneumatic sinuses. All skulls are oriented in their 'alert' postures, determined by orienting the skull such that the lateral semicircular canal is horizontal. Scale bars equal 10 cm, except that for *B. virginianus* which equals 2 cm.

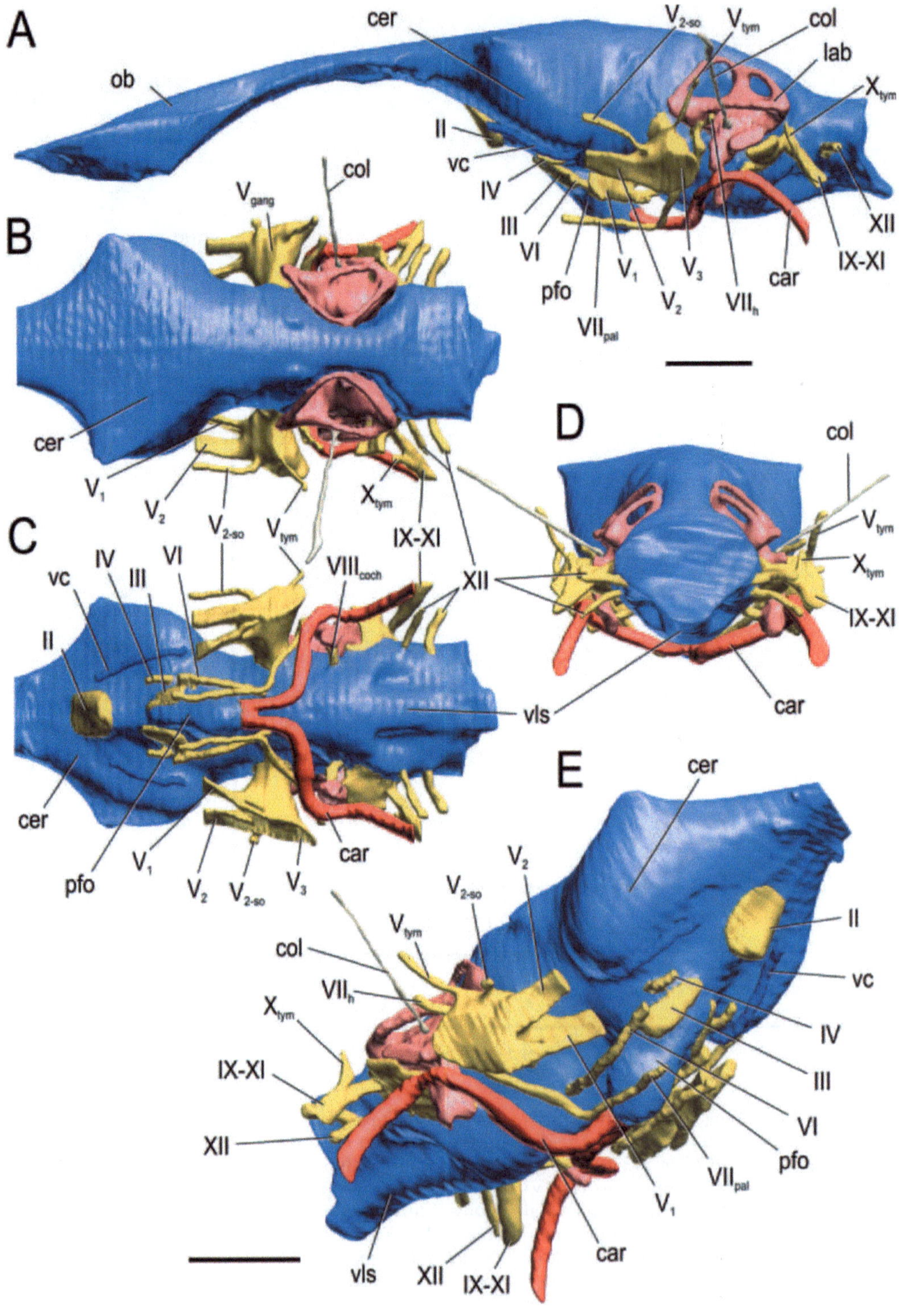

FIG. 6.3. Cranial endocast, endosseous labyrinth, and some endocranial vascular structures of an Australian freshwater crocodile, *Crocodylus johnstoni* (OUVC 10425), derived from surface renderings of CT scan data. **A**, left lateral view. **B**, dorsal view, with olfactory tract truncated. **C**, ventral view, with olfactory tract truncated. **D**, caudal view. **E**, right rostroventrolateral view. Color scheme: cranial endocast, *blue*; endosseous labyrinth, *pink*; nerve canals (most of which also transmit veins), *yellow*; smaller venous canals, *dark blue*; arterial canals, *red*; columella, *pale yellow*. Scale bars equal 1 cm.

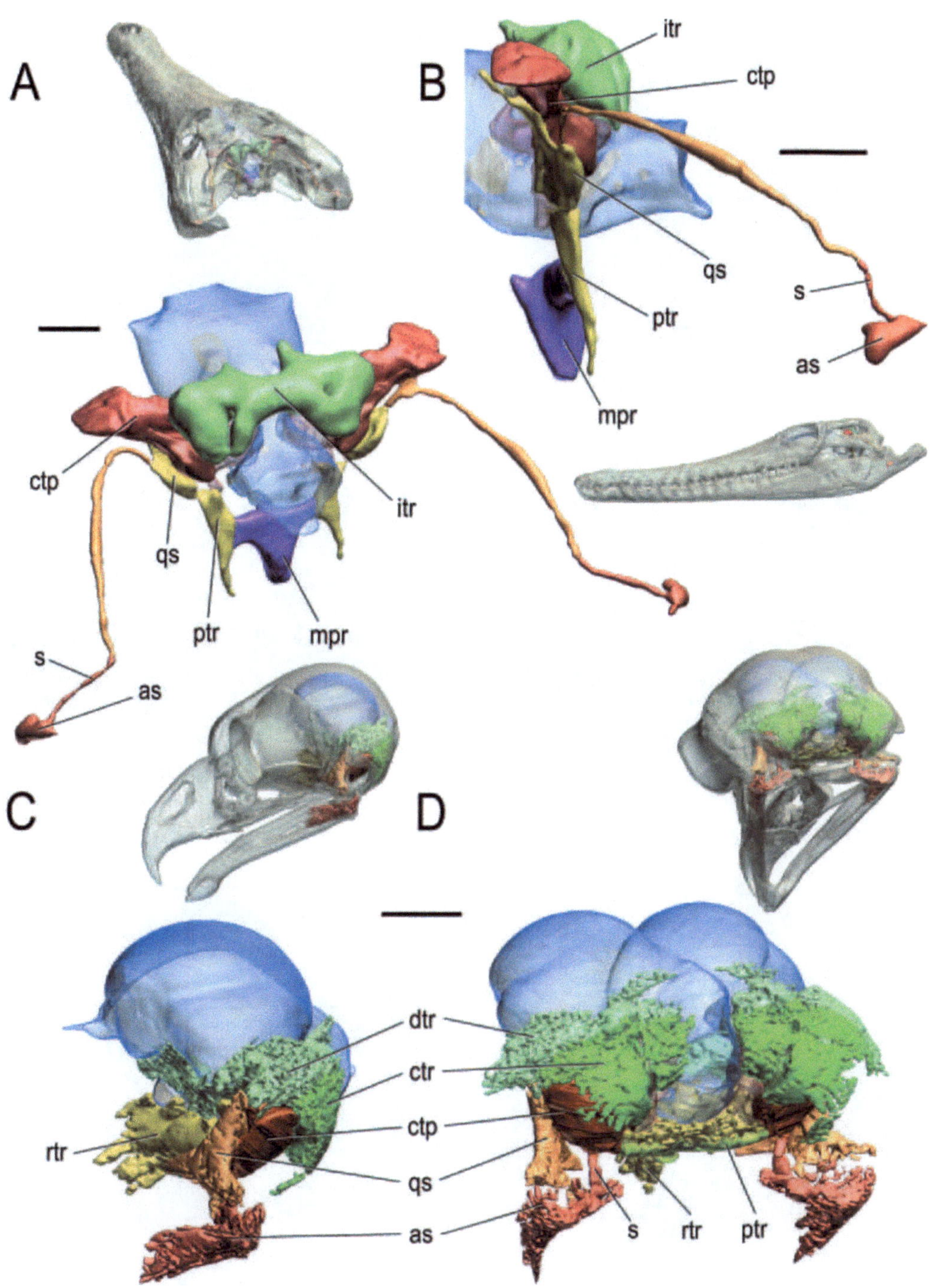

FIG. 6.6. Pneumatic sinuses of the braincase region of, (**A–B**), an Australian freshwater crocodile, *Crocodylus johnstoni* (OUVC 10425), and (**C–D**), a great horned owl, *Bubo virginianus* (OUVC 10220), viewed surrounding a semitransparent cranial endocast, derived from surface renderings of CT scan data. Thumbnail views of the whole skull are provided above the main image (except B, where it is below) to show the orientation of the main image. (**A**) *C. johnstoni* in left caudodorsolateral view. (**B**) *C. johnstoni* in left lateral view. (**C**) *B. virginianus* in left lateral view. (**D**) *B. virginianus* in left caudoventrolateral view. Scale bars equal 1 cm.

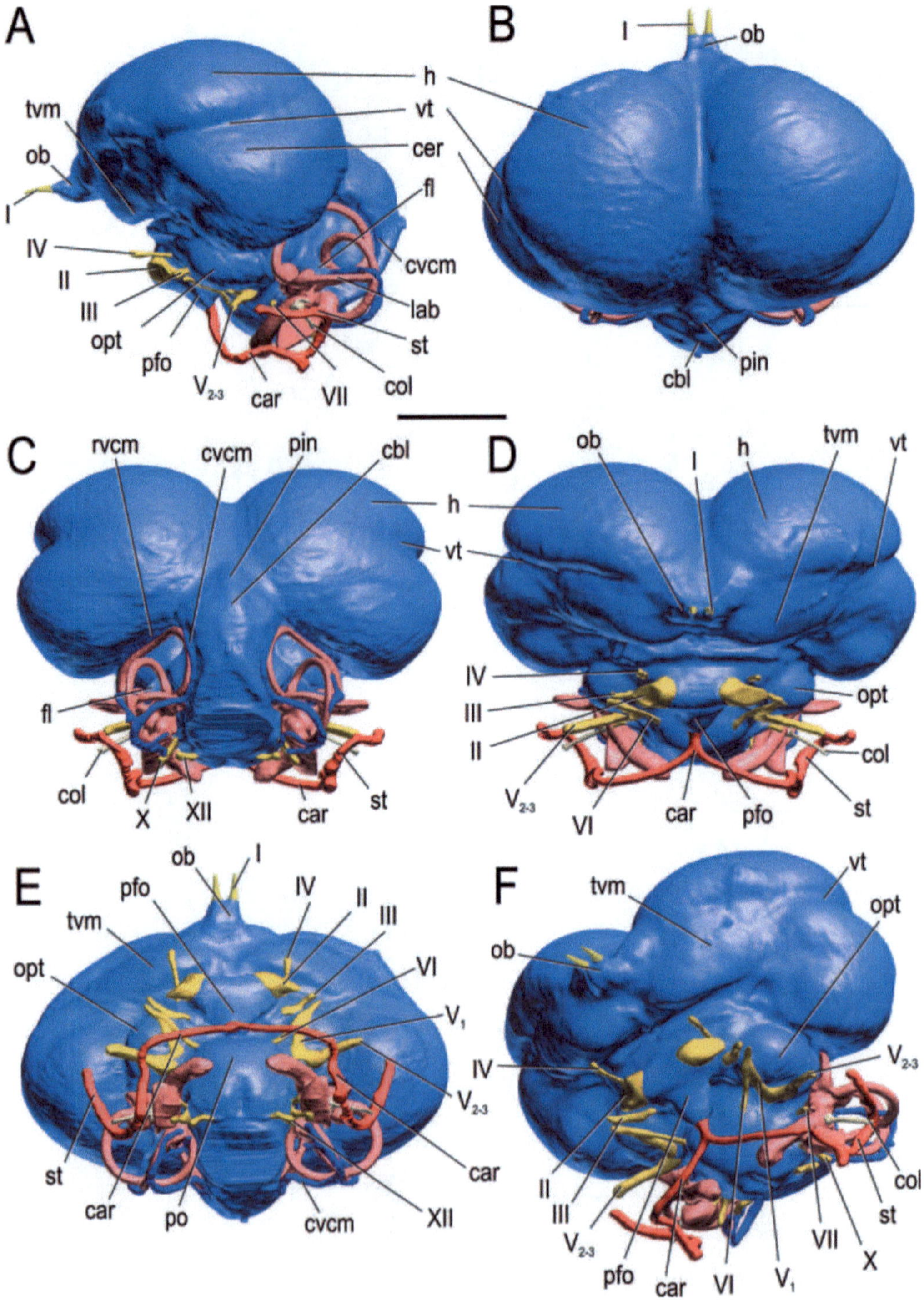

FIG. 6.7. Cranial endocast, endosseous labyrinth, and some endocranial vascular structures of a great horned owl, *Bubo virginianus* (OUVC 10220), derived from surface renderings of CT scan data. **A**, left lateral view. **B**, dorsal view. **C**, caudal view. **D**, rostral view. **E**, ventral view. **F**, left rostroventrolateral view. Color scheme: cranial endocast, *blue*; endosseous labyrinth, *pink*; nerve canals (most of which also transmit veins), *yellow*; smaller venous canals, *dark blue*; arterial canals, *red*; columella, *pale yellow*. Scale bar equals 1 cm.

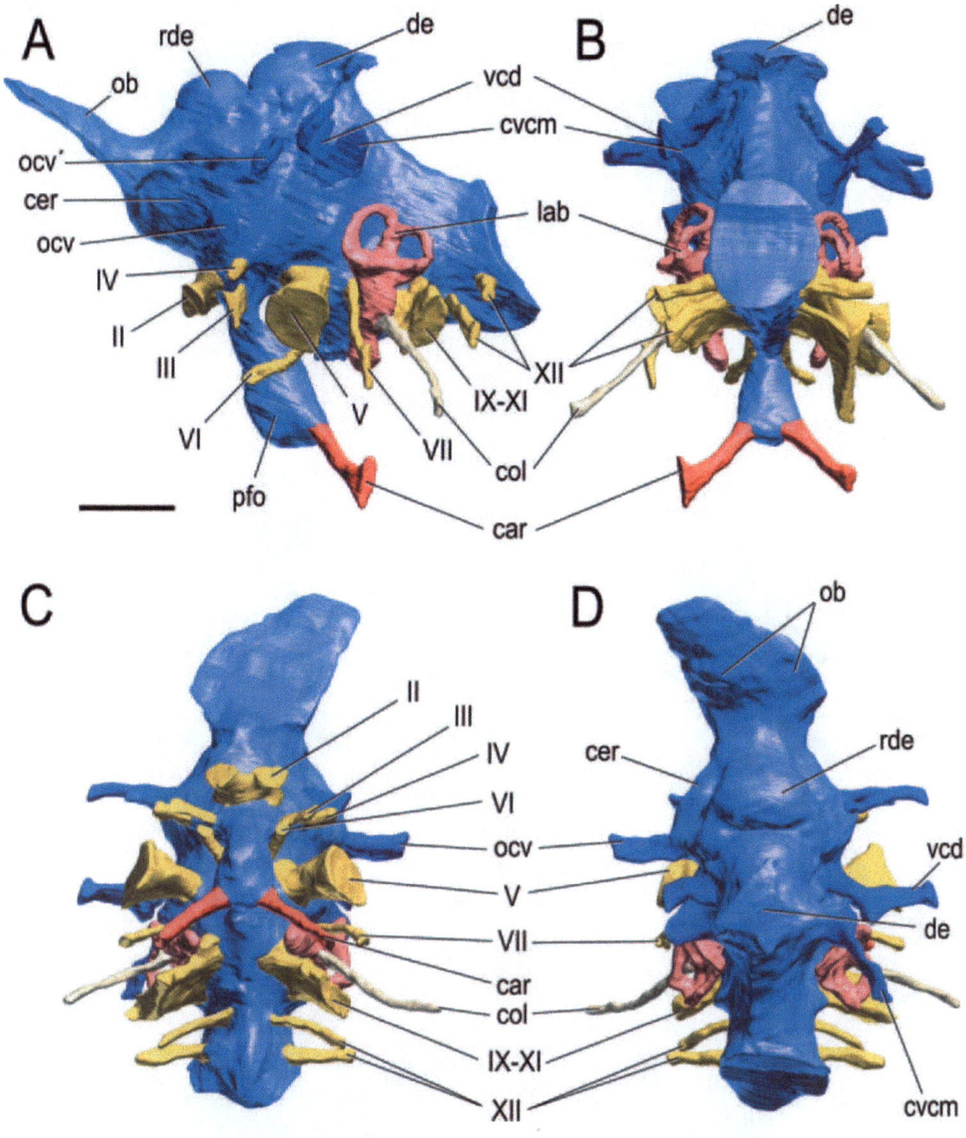

FIG. 6.8. Cranial endocast, endosseous labyrinth, and some endocranial vascular structures of the sauropod dinosaur, *Camarasaurus lentus* (CM 11338), derived from surface renderings of CT scan data. **A**, left lateral view. **B**, caudal view. **C**, ventral view. **D**, dorsal view. Color scheme: cranial endocast, *blue*; endosseous labyrinth, *pink*; nerve canals (most of which also transmit veins), *yellow*; smaller venous canals, *dark blue*; arterial canals, *red*; columella, *pale yellow*. Scale bar equals 2 cm.

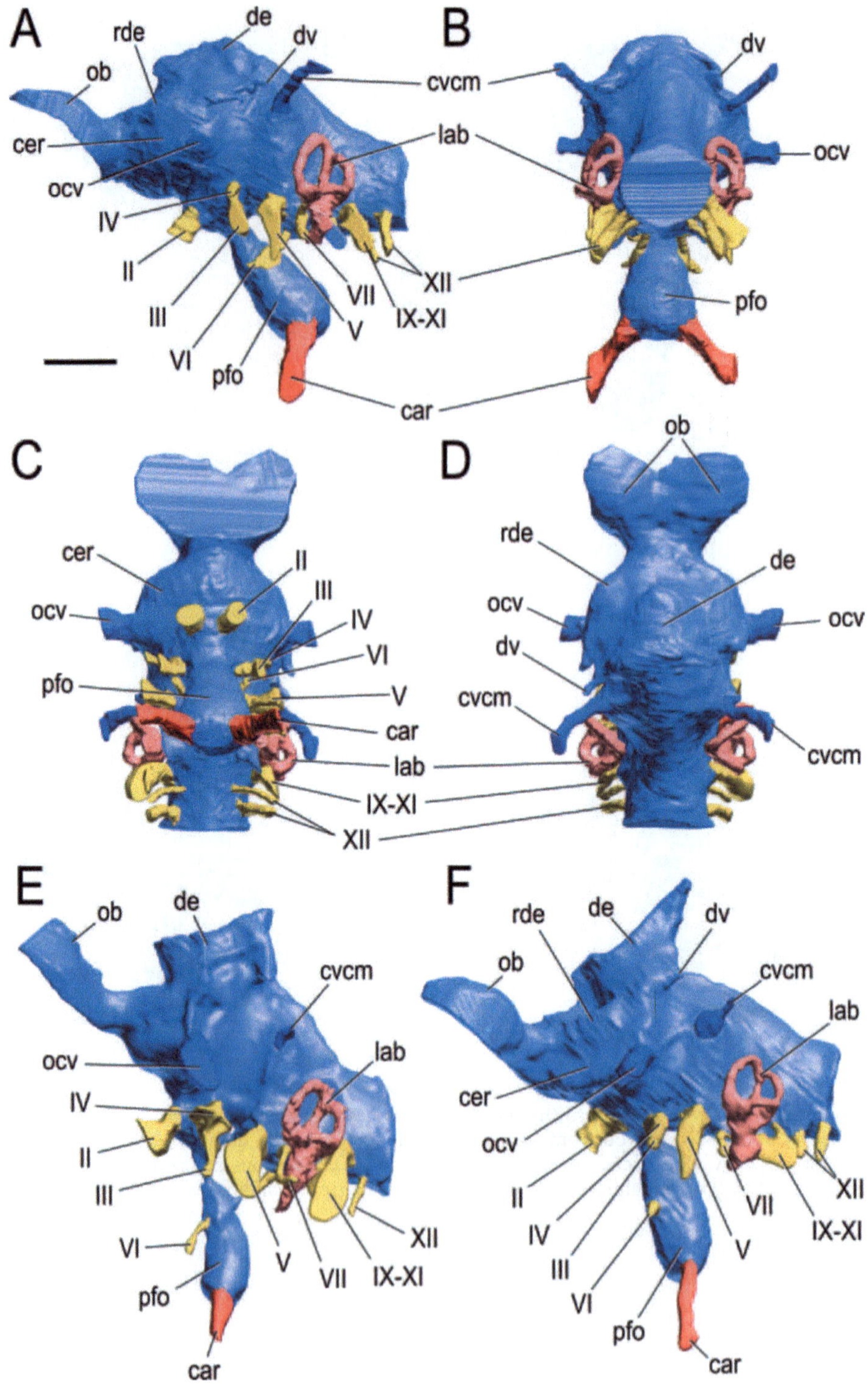

FIG. 6.9. Cranial endocast, endosseous labyrinth, and some endocranial vascular structures of the sauropod dinosaur, *Diplodocus longus*, derived from surface renderings of CT scan data. **A–D**, CM 3452. **E**, AMNH 694. **F**, CM 11161. **A**, **E**, **F**, left lateral view. **B**, caudal view. **C**, ventral view. **D**, dorsal view. Color scheme: cranial endocast, *blue*; endosseous labyrinth, *pink*; nerve canals (most of which also transmit veins), *yellow*; smaller venous canals, *dark blue*; arterial canals, *red*. Scale bar equals 2 cm.

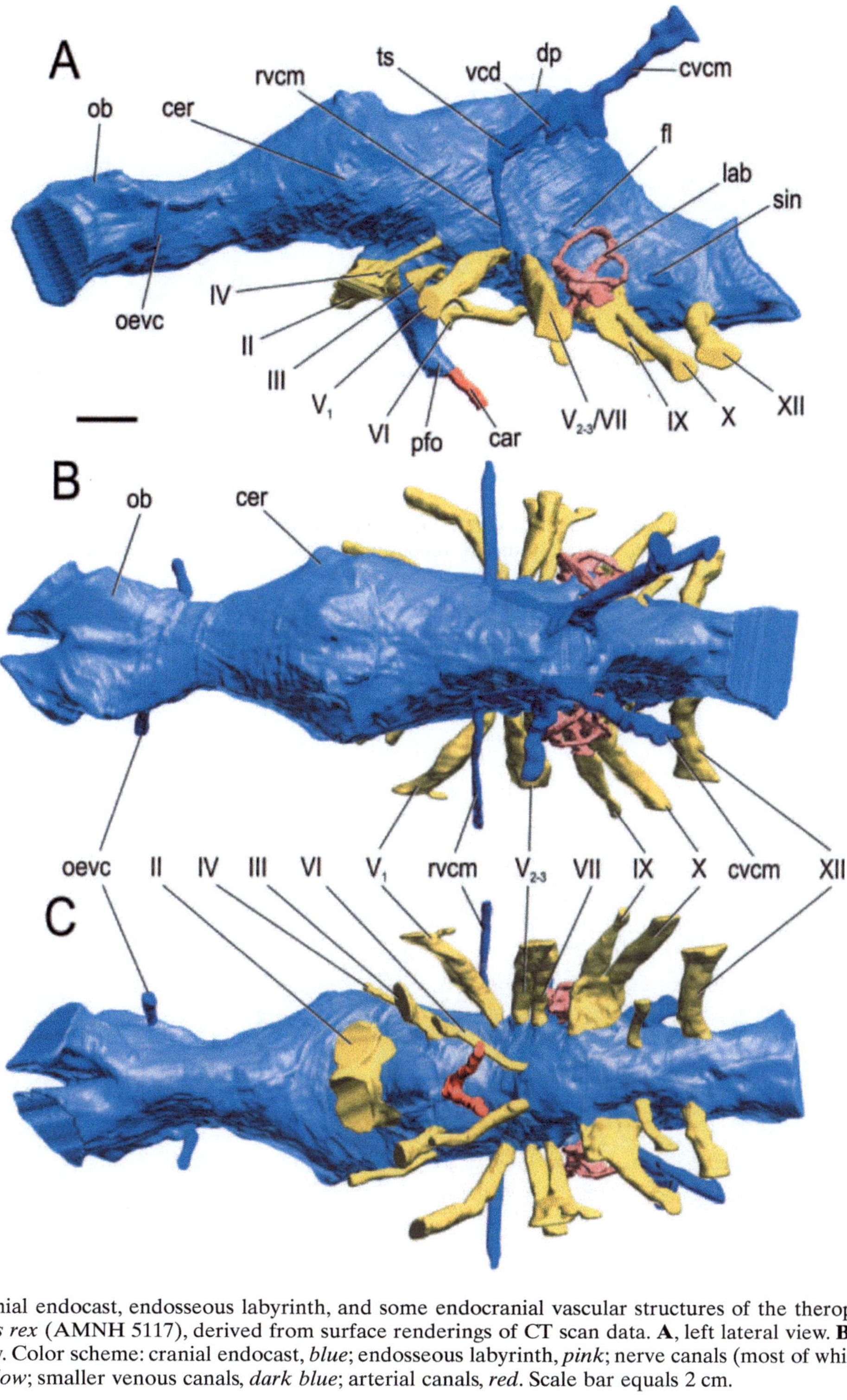

FIG. 6.10. Cranial endocast, endosseous labyrinth, and some endocranial vascular structures of the theropod dinosaur, *Tyrannosaurus rex* (AMNH 5117), derived from surface renderings of CT scan data. **A**, left lateral view. **B**, dorsal view. **C**, ventral view. Color scheme: cranial endocast, *blue*; endosseous labyrinth, *pink*; nerve canals (most of which also transmit veins), *yellow*; smaller venous canals, *dark blue*; arterial canals, *red*. Scale bar equals 2 cm.

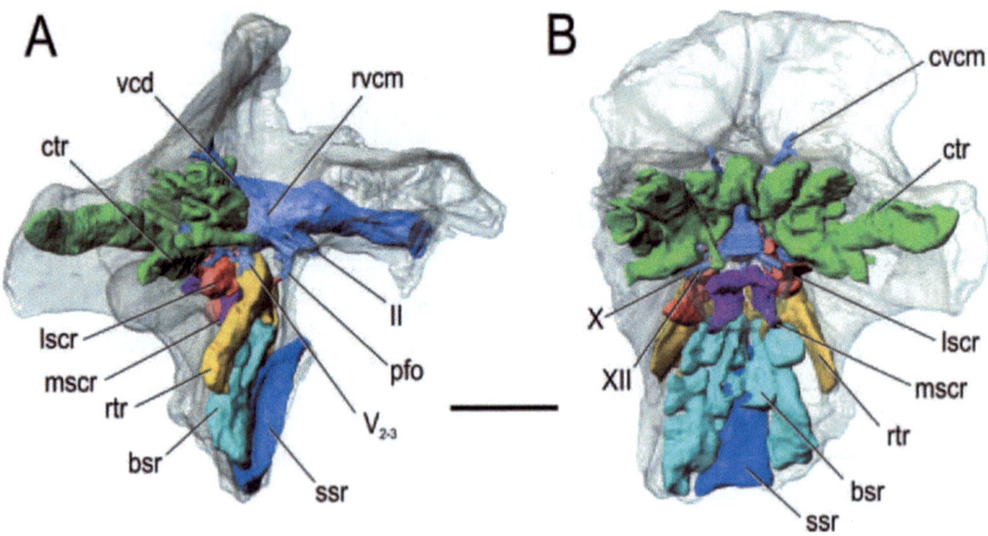

FIG. 6.11. Pneumatic sinuses and cranial endocast within the semitransparent bony braincase of the theropod dinosaur, *Tyrannosaurus rex* (AMNH 5117), derived from surface renderings of CT scan data. **A**, right lateral view. **B**, caudal view. Scale bar equals 10 cm.

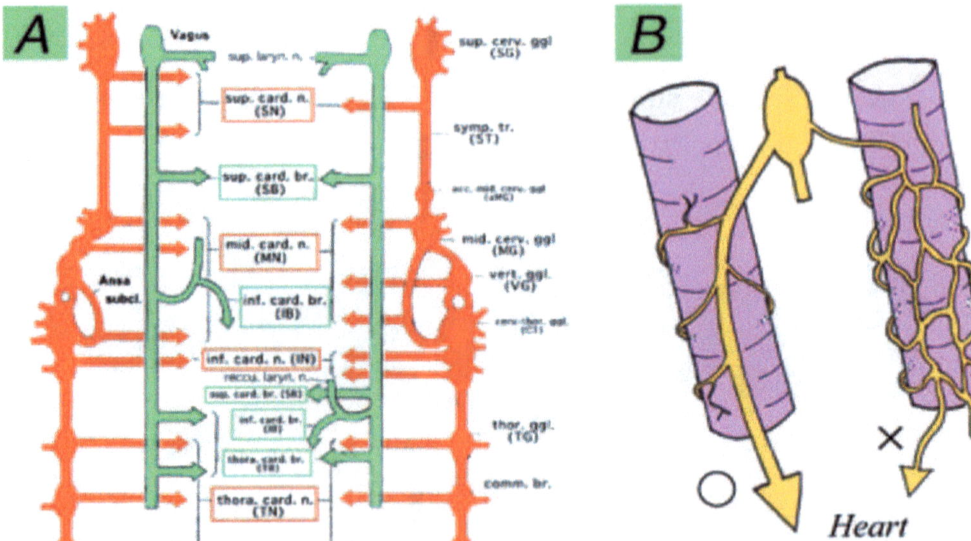

FIG. 7.1. Diagrams showing our definition of the autonomic cardiac nervous system. (**A**) Each cardiac nerve/branch is named according to its origin. The sympathetic cardiac nerves and the parasympathetic vagal cardiac branches are colored in *orange* and *green*, respectively. (**B**) The cardiac nerves/branches are regarded as nerves with direct connections via the cardiac plexus, except for slender thin nerves/branches like the sensory nerves around great vessels.

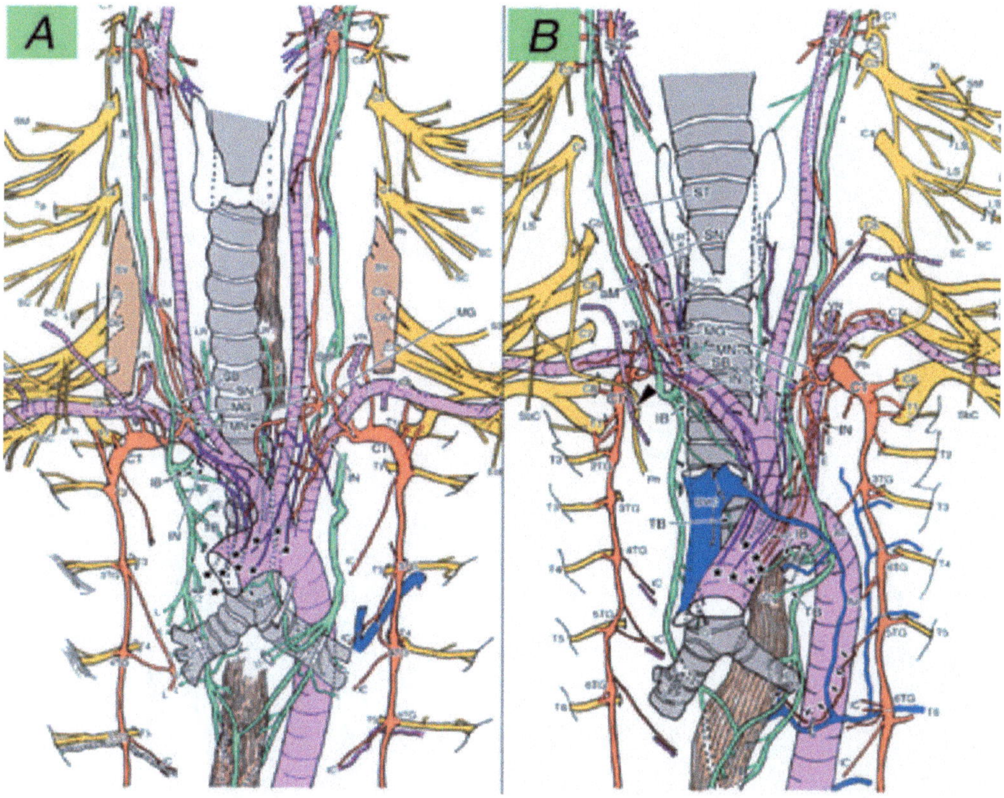

FIG. 7.2. Diagram showing the autonomic cardiac nervous system and its surrounding structure in Asian Old World monkeys *(Cercopithecidae)*. (**A**) rhesus monkey *(Macaca mulatta)*. (**B**) stump-tailed monkey (*Macaca arctoides*), ventral view. The black and white stars show the cardiac nerves/branches entering from the arterial and venous parts of the pericardium reflection, respectively (After Kawashima et al. 2005). aM, accessory middle ganglion; CT, cervicothoracic (stellate) ganglion; IB, inferior (vagal) cardiac branch; IN, inferior cervical cardiac nerve; L, nerve branches to lung; LR, recurrent laryngeal nerve of vagus nerve; MG, middle cervical ganglion; MN, middle cardiac nerve; Ph, phrenic nerve; SVC, superior vena cava; SB, superior (vagal) cardiac branch; SbC, nerve to subclavian muscle; SC, supraclavicular nerve; SG, superior cervical ganglion; SM, nerve branch to sternocleidomastoid muscle; SN, superior cardiac nerve; SS, suprascapular nerve; ST, sympathetic trunk; TB, thoracic (vagal) cardiac branch; Tp, nerve branch to trapezius muscle; VN, vertebral nerve; *X*, vagus nerve; *XI*, accessory nerve; 2-6TG, second to sixth thoracic ganglia.

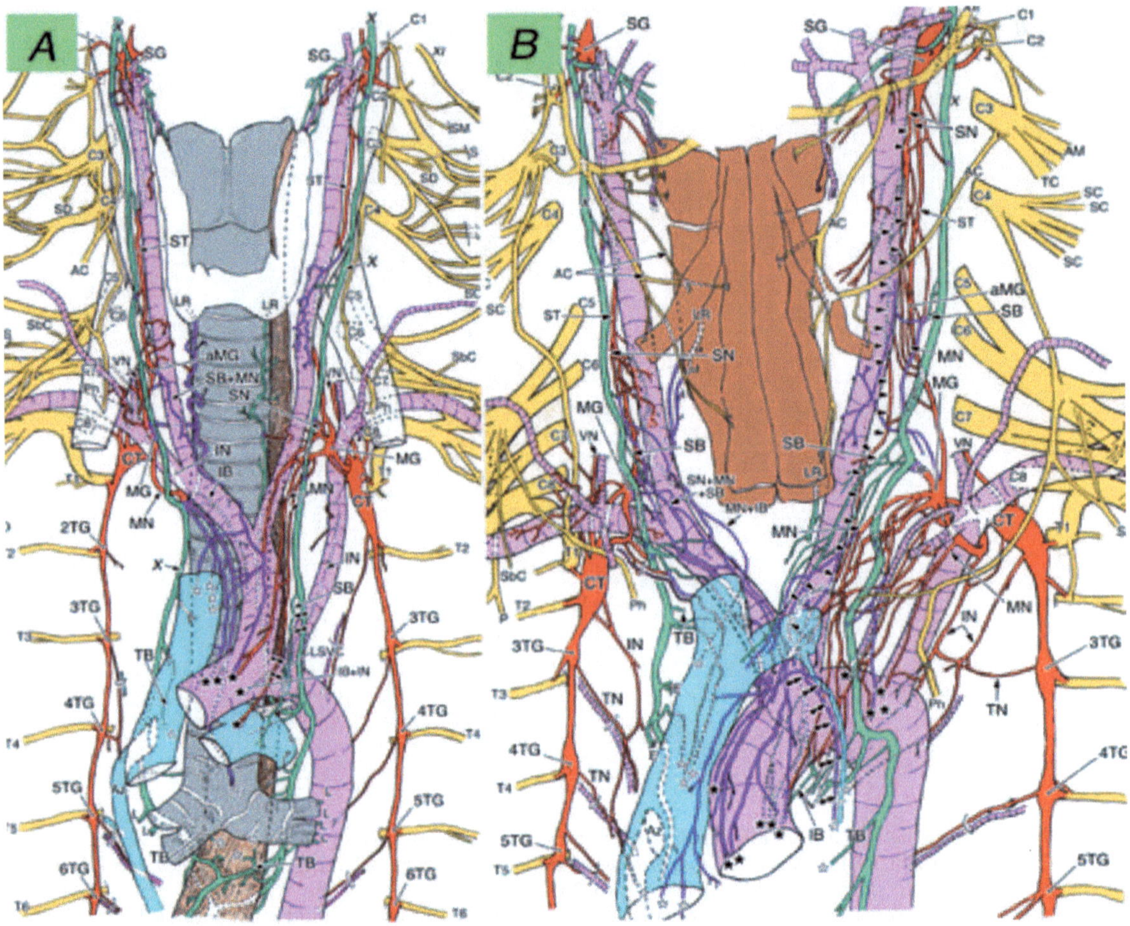

FIG. 7.3. Diagrams showing the autonomic cardiac nervous systems in African Old World monkeys *(Cercopithecidae)*. (**A**) Savanna monkey *(Chlorocebus aethiops)*. (**B**) Anubis baboon *(Papio anubis)*. Viewed from the ventral aspect. The black and white stars show the cardiac nerves/branches entering from the arterial and venous parts of the pericardium reflection, respectively (After Kawashima et al. 2007). AC, cervical ansa; AM, greater auricular nerve; aMG, accessory middle ganglion; CT, cervicothoracic (stellate) ganglion; IB, inferior (vagal) cardiac branch; IN, inferior cervical cardiac nerve; L, nerve branches to lung; LR, recurrent laryngeal nerve of vagus nerve; MG, middle cervical ganglion; MN, middle cardiac nerve; P, pectoral nerve; Ph, phrenic nerve; SB, superior (vagal) cardiac branch; SbC, nerve to subclavian muscle; SC, supraclavicular nerve; SD, dorsal scapular nerve; SG, superior cervical ganglion; SM, nerve branch to sternocleidomastoid muscle; SN, superior cardiac nerve; ST, sympathetic trunk; TB, thoracic (vagal) cardiac branch; TC, transverse cervical nerve; TN, thoracic cardiac nerve; VN, vertebral nerve; *X*, vagus nerve; *XI*, accessory nerve; *XII*, hypogastric nerve; 2-6TG, second to sixth thoracic ganglia.

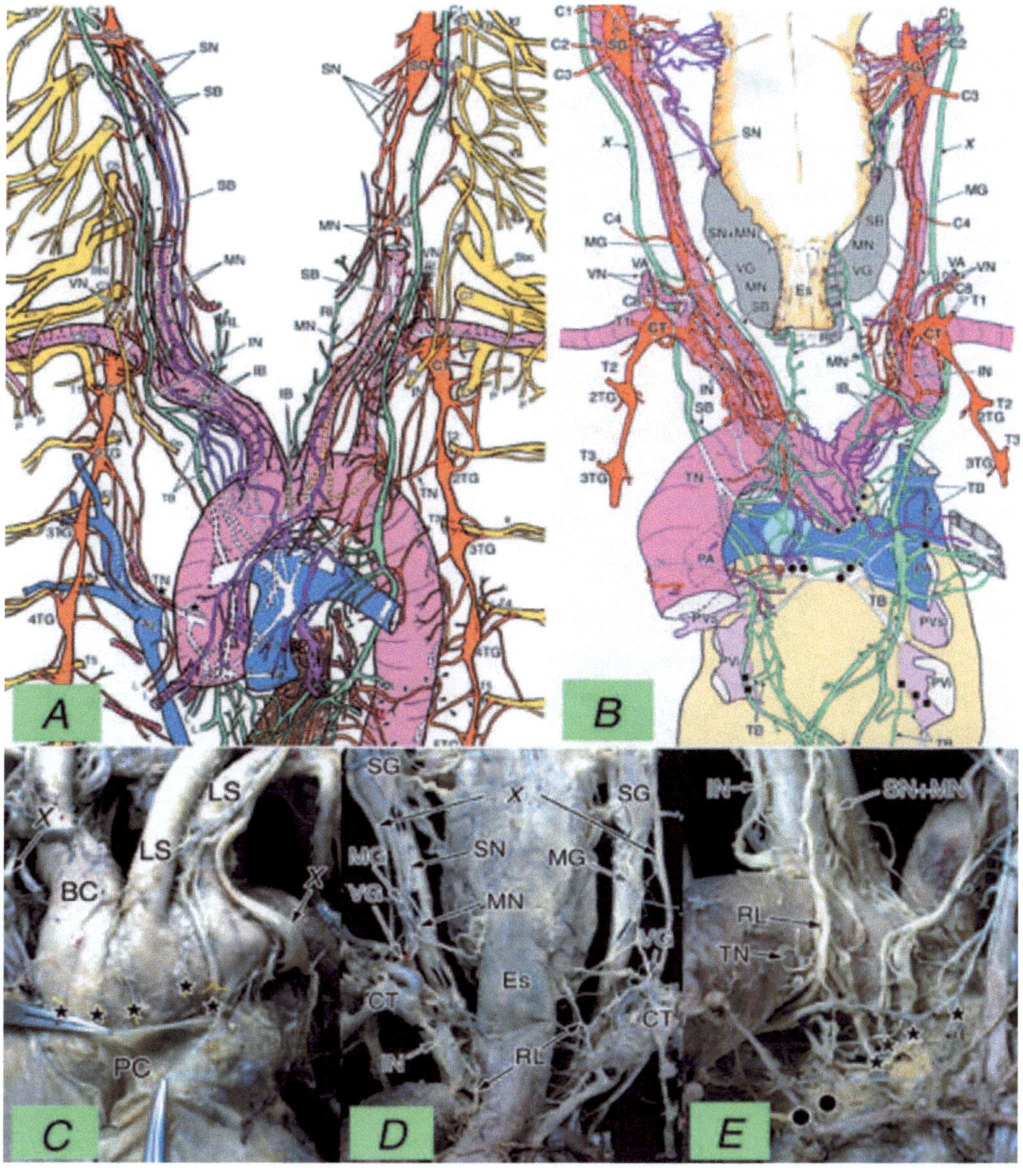

FIG. 7.4. Diagrams and photographs showing the autonomic cardiac nervous systems (ACNS) in humans. (**A**) Ventral view. (**B**) Dorsal view. The sympathetic cardiac nerves, vagal cardiac branches, and the mixture nerves (cardiac plexus) are colored orange, green, and purple, respectively. (**C**) Ventral view. The reflection of the pericardium in the arterial part of the *hilum cordis* and its nerve entrances are clearly observed with incision of a part of the parietal pericardium and its retraction by forceps. (**D**) Dorsal view. The ACNS accompanying the great arteries. (**E**) The complex cardiac plexus. An enlargement of Fig. 7.1B. As seen in this case, all cardiac nerves and branches are normally observed along the great vessels. Black stars (★), squares (■), and circles (•) indicate the inlet/outlet of the cardiac nerves/branches from the arterial, venous parts of, and the *hilum cordis* (After Kawashima 2005; Kawashima and Sasaki 2006). CT, cervicothoracic (stellate) ganglion; Es, esophagus; IB, inferior (vagal) cardiac branch; IN, inferior cervical cardiac nerve; LC, left common carotid artery; LS, left subclavian artery; MG, middle cervical ganglion; MN, middle cardiac nerve; PA, pulmonary artery; PC, pericardial cavity; PVi, inferior pulmonary vein; PVs, superior pulmonary vein; RL, recurrent laryngeal nerve; SB, superior (vagal) cardiac branch; SbC, branch to the subclavian muscle; SG, superior cervical ganglion; SN, superior cardiac nerve; SVC, superior vena cava; TB, thoracic (vagal) cardiac branch; TG, thoracic ganglia; TN, thoracic cardiac nerve; VA, vertebral artery; VG, vertebral ganglion; VN, vertebral nerve; *X*, vagus nerve.

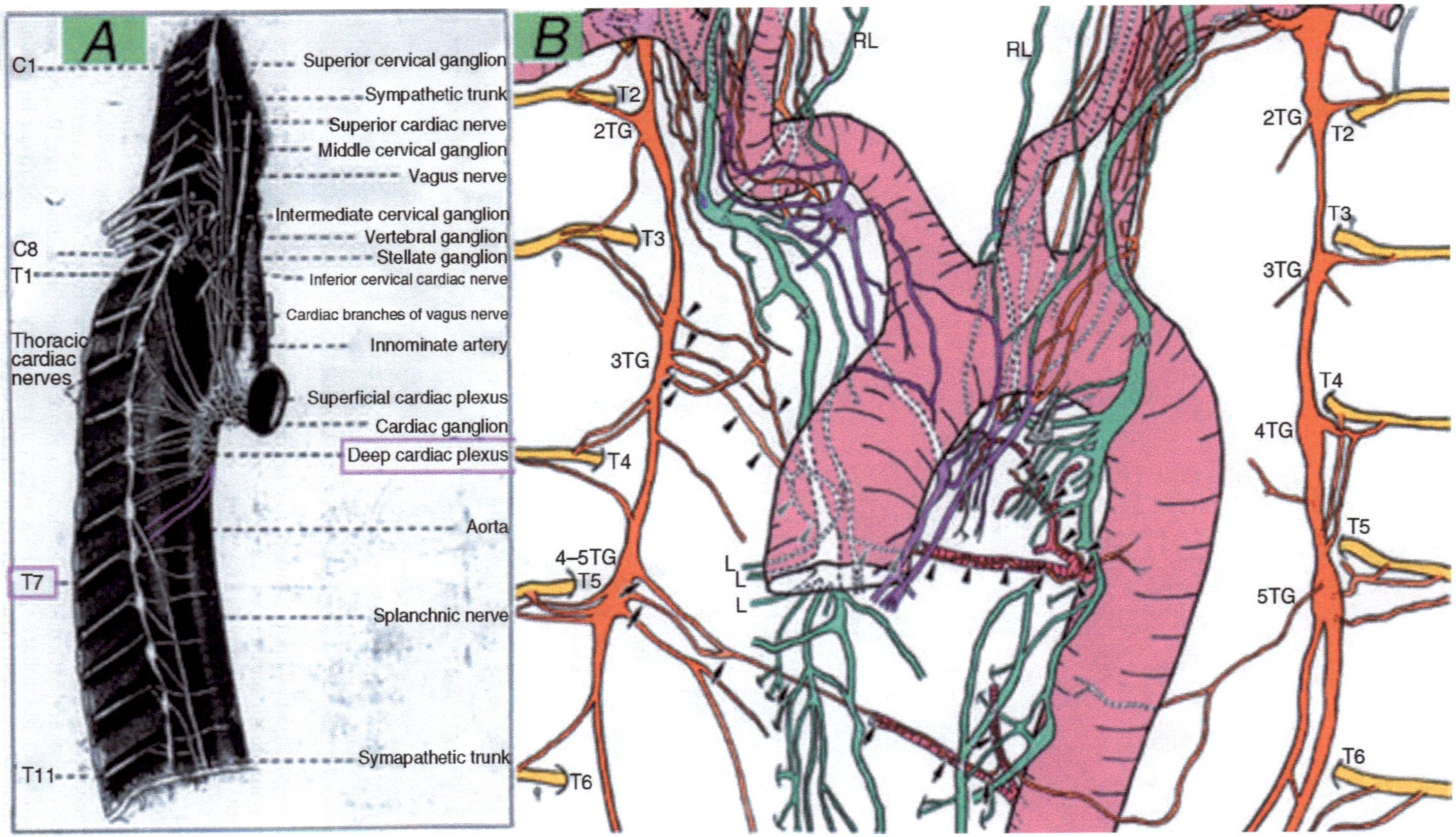

FIG. 7.5. Different descriptions of the course in the right thoracic cardiac nerve. (**A**) conventional description (Saccomanno 1943). The right thoracic cardiac nerve originating from the lower ganglia, such as the fifth to seventh thoracic ganglia, ascends to reach the heart at a position higher than the origin of the thoracic cardiac nerve, regardless of the course of the intercostal vessels. (**B**) Our new description (recurrent right thoracic cardiac nerve: Kawashima 2005; Kawashima and Sasaki 2005). It descends obliquely along the intercostal vessels. After arriving at the thoracic aorta, it turns there and ascends along the aorta, finally distributing to the heart via the left venous part of cardiac hilum or connecting to the cardiac plexus as shown by the arrow heads.

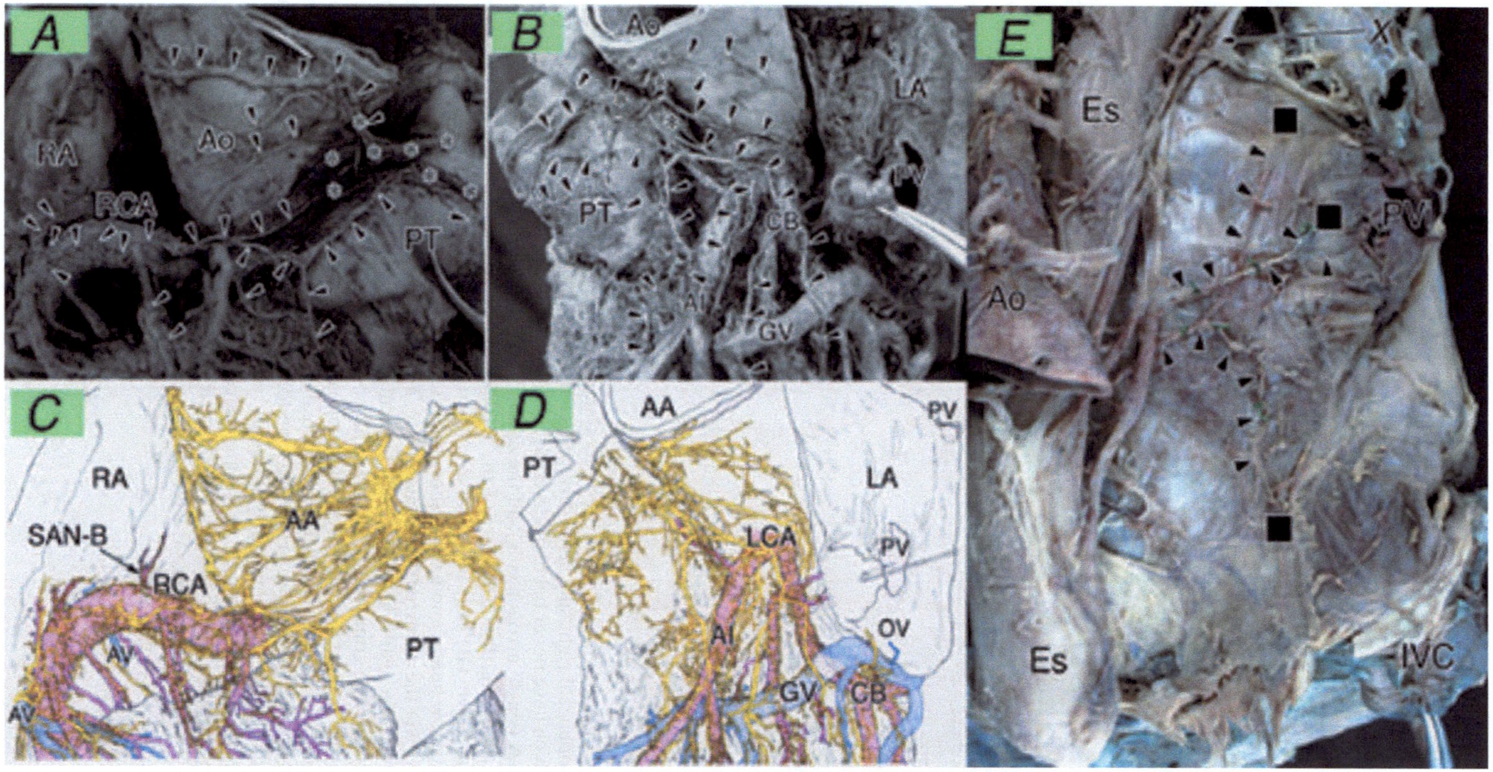

FIG. 7.6. Photographs and diagrams showing the intrinsic peripheral cardiac distributions. (**A**, **C**) Arterial group consisting of relatively thick cardiac nerves/branches. Right sides. (**B**, **D**) Left side. The bundle (*) of the cardiac plexus courses between the aorta and the pulmonary trunk, runs along the great vessels and both coronary arteries, and then distributes to the atria, ventricles, and coronary arteries. (**E**) Venous group consisting of very thin branches, dorsal view. Arrow heads and squares (■) show the course and entrance to the pericardium, respectively (After Kawashima 2005). AI, anterior interventricular branch of left coronary artery; CB, circumflex branch of left coronary artery; Es, esophagus; GV, great cardiac vein; IVC, inferior vena cava; LA, left atrium; LCA, left coronary artery; PV, pulmonary vein; RA, right atrium; RCA, right coronary artery; X, vagus nerve.

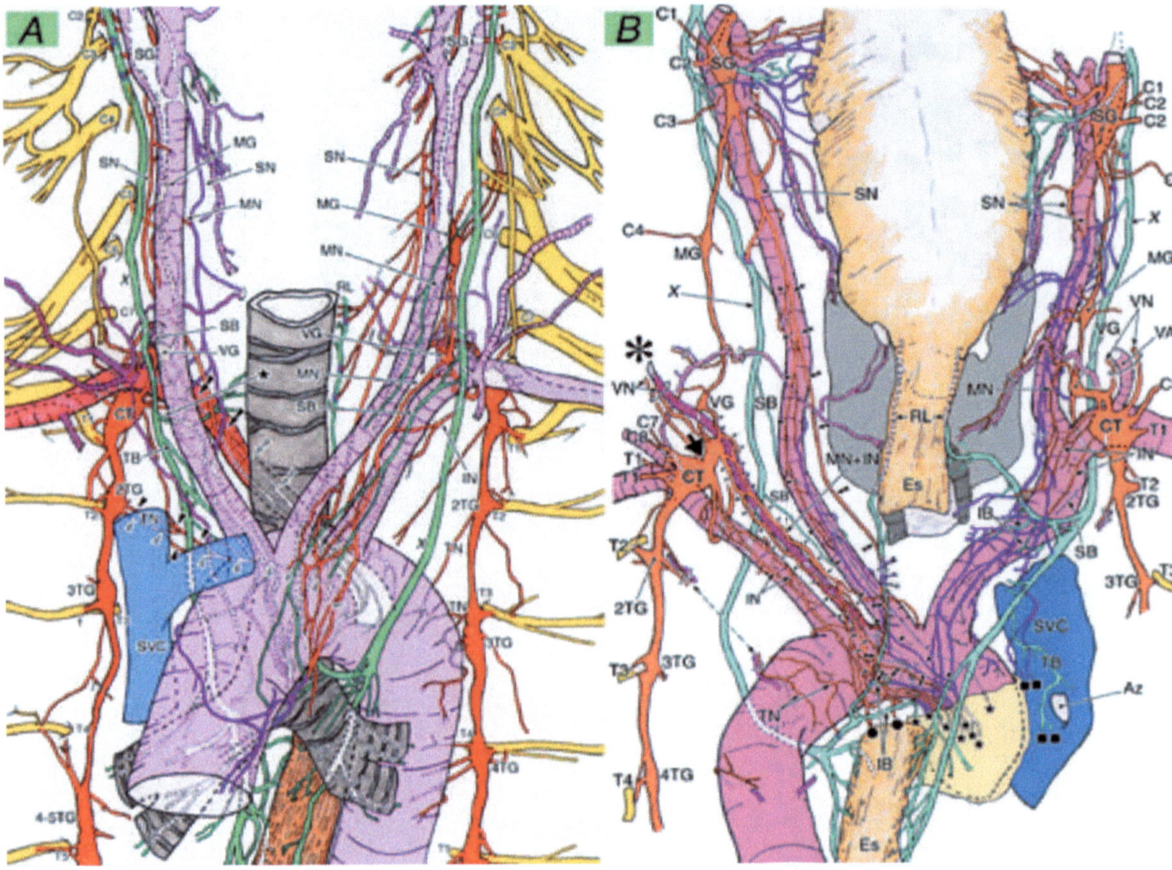

FIG. 7.8. Topological changes of the autonomic cardiac nervous system (ACNS) in human individuals with anomalies of branchial arteries. (**A**) Retroesophageal right subclavian artery (RERSA) viewed from the ventral aspect. Although the right thoracic cardiac nerves, which have never been observed to accompany the normal right subclavian artery, ran along the proximal part of the RERSA (single arrow heads), all the cardiac nerves were recognized. Double arrow heads and black star show the cardiac nerve along the distal part of the RERSA and conventional finding of the corresponding nerve to the recurrent laryngeal nerve. (**B**) Left vertebral artery originating directly from the aortic arch, dorsal view. The single arrow heads, double arrow heads, and arrows show the direct distribution of the superior cardiac nerve to the heart, the superior cardiac nerve via the recurrent laryngeal nerve, and the inferior cardiac nerve along the anomalous left vertebral artery, respectively. These results in two types of branchial arterial anomalies suggest that additional cardiac nerves along the anomalous artery were found in addition to the normal ACNS composition although the ACNS shift have been considered when a great arterial branching anomaly appeared (After Kawashima and Sasaki 2005, 2007). CT, cervicothoracic (stellate) ganglion; Es, esophagus; IB, inferior (vagal) cardiac branch; IN, inferior cervical cardiac nerve; LC, left common carotid artery; LS, left subclavian artery; MG, middle cervical ganglion; MN, middle cardiac nerve; PA, pulmonary artery; PC, pericardial cavity; PVi, inferior pulmonary vein; PVs, superior pulmonary vein; RL, recurrent laryngeal nerve; SB, superior (vagal) cardiac branch; SG, superior cervical ganglion; SN, superior cardiac nerve; SVC, superior vena cava; TB, thoracic (vagal) cardiac branch; TG, thoracic ganglia; TN, thoracic cardiac nerve; VA, vertebral artery; VG, vertebral ganglion; VN, vertebral nerve; *X*, vagus nerve.

1 Head Anatomy of Male and Female Mongolian Gazelle – A Striking Example of Sexual Dimorphism

Roland Frey[1], Alban Gebler[2], Kirk A. Olson[3], Daria Odonkhuu[3], Guido Fritsch[4], Nyamsuren Batsaikhan[5], and Ingo W. Stuermer[6]

1.1 Introduction

The first scientific description of the Mongolian gazelle by Pallas (1777) already depicted a sexual dimorphism of the larynx and the thyrohyoid connection (Fig. 1.1).

Compared to the vocal tract in females and to other mammalian species, the larynx of male Mongolian gazelle has been evolutionarily enlarged considerably and shows several newly evolved specializations, e.g. a paired lateral ventricle and an unpaired palatinal pharyngeal pouch (PPP – cf. Kleinschmidt 1961; Frey and Riede 2003; Frey and Gebler 2003).

The rut of Mongolian gazelle is in December. Dominant males are almost permanently producing loud head-up rutting barks when rounding up their harem females. Simultaneous with each bark, the larynx is pronouncedly retracted (Frey et al. 2008). Obviously, an enlarged and specialized larynx contributes to the acoustic display, which is an integral part of male courtship behaviour.

Apart from the acoustic display, the neck region of the male is visually conspicuous by a pronounced laryngeal prominence that is further accentuated by a specific colour pattern consisting of a rostral light-coloured bib and lateral dark-coloured areas (Fig. 1.2). This colour pattern of the male neck region disappears after the rut as the dark guard hairs fall out gradually.

This chapter investigates the head anatomy of male and female Mongolian gazelle focussing on structures of the vocal tract, particularly the hyoid apparatus, the pharynx and the larynx. An integrated approach – modern and classic combined – was applied. First, 3D virtual reconstructions based upon computed tomographic scans of undamaged specimens were made. Second, specimens were submersed in water and dissected layer by layer following the classic techniques of comparative anatomy.

Computed tomographic investigations have the great advantage of being non-invasive. Basically, they can be carried out on living (anaesthesized) individuals or on dead specimens without any damaging. A further advantage is the speed of the investigation: scanning itself lasts only some seconds and, after loading the raw scans to a work station, investigations on 3D virtual reconstructions of the specimens can begin rapidly. Disadvantages are the enormous purchase and running costs of the machine and the necessity for a specifically equipped room (protection from X-rays etc.). An additional disadvantage is the limited scan field (about 50 cm in diameter) as current computer tomographs have been designed for medical investigations of humans. As a consequence, there is a limit to the scanning of large or entire individuals.

In contrast, investigations by classic dissection are invasive and slow: they can be carried out on dead specimens only and the successive removing of layers requires considerable manual skill and experience of the investigator and a considerable

[1]Leibniz Institute for Zoo and Wildlife Research (IZW), Berlin, Germany, [2]Institute for Theoretical Biology, Humboldt University, Berlin, Germany, [3]Wildlife Conservation Society (WCS), Mongolia Program Office, Ulaanbaatar, Mongolia, [4]Clinic for small animals, Oertzenweg, Berlin, Germany, [5]Faculty of Biology, Department of Zoology, National University of Mongolia, Ulaanbaatar, Mongolia, [6]Sensory Developmental Research Group, Department of Phoniatrics and Pedaudiology, Georg-August-University, Goettingen, Germany

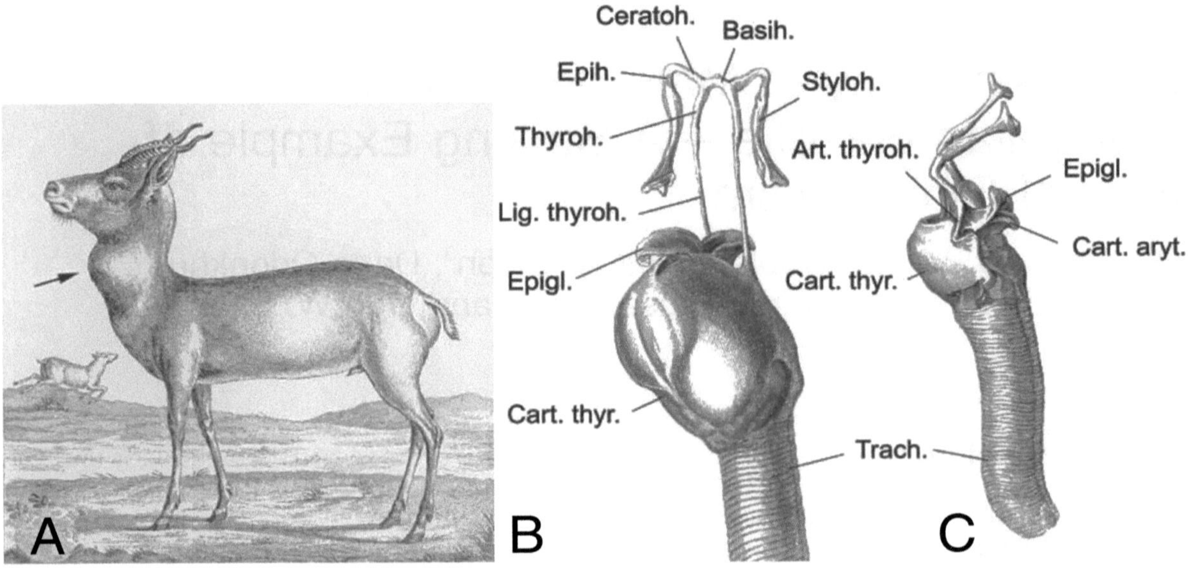

FIG. 1.1. Habitus of male Mongolian gazelle (**A**) with the characteristic laryngeal prominence (*arrow*). Female in the background without horns and laryngeal prominence. Sexual dimorphism of hyoid apparatus and larynx of male (**B**) and female (**C**) Mongolian gazelle. In the rutting male the thyrohyoid connection is transformed into a ligament (Lig. thyroh.) whereas the female retains the primitive short articulation (Art. thyroh.). (**A**, **B** and **C** after Pallas 1777)

FIG. 1.2. Habitus and diagrammatic colour pattern of rutting male Mongolian gazelle according to video single frames (see insets). Laterofrontal view (**A**) and lateral view (**B**) evidencing the laryngeal prominence and the accentuation of the ventral neck region (see Color Plates, Fig. 1.2).

amount of time (days, weeks). However, these disadvantages become outweighed by the higher 'resolution' of classic dissection allowing for documentation of structures that cannot be adequately documented by modern imaging techniques (muscles, nerves etc.). Additional advantages of classic dissection are the low costs and the absence of a size limit of the specimens.

Third, a synthesis brought together results obtained by both methods, additionally included behavioural and acoustical data and attempted a preliminary morphofunctional analysis.

1.2 The Larynx

In addition to more obvious sexual dimorphisms as, e.g., horns, male Mongolian gazelles have evolved a larynx about double the size that would be predicted from male-to-female body mass proportions. The male larynx achieves a weight of about 600 g and expands the ventral neck region considerably so that it contributes to the sexual dimorphism of male and female external appearance (Fig. 1.2). By contrast, the relative size of the female larynx corresponds to that of other species of the bovid family, as, e.g., sheep and goat. Sexually dimorphic larynx size is also revealed by computed 3D reconstructions of skeletal elements in the head and neck region (Fig. 1.3).

Apart from a size difference, the male larynx has evolved several unique features that are not found in the female, e.g. an exceptionally large epiglottis, a broad ventral keel of the thyroid cartilage, a paired lateral laryngeal ventricle and a voluminous vocal fold supported by a tough, cymbal-shaped fibroelastic pad.

The laryngeal cartilages, including the epiglottis, can be analysed in virtual 3D reconstructions based upon serial computed tomographic scans.

By contrast to the typically-sized, small epiglottis of the female, the exceptionally large size of the epiglottis in male Mongolian gazelles may result from the mobility of their larynx that has evolved in the context of acoustic display. Videos taken in the natural habitat reveal that the larynx of adult males descends to a low mid-neck resting position during the rut in winter. Despite the large size of the epiglottis, this entails a loss of contact between soft palate and epiglottis and the danger of swallowing the wrong way. The larynx in adult males is further retracted intermittently from this low resting position down to the thoracic aperture while they produce their characteristic rutting barks. The protraction of the larynx from its low resting position towards the lower jaw will, however, enable the particularly large epiglottis of adult male Mongolian gazelle to regain contact with the soft palate in the course of swallowing and, thus, ensure a safe food passage into the oesophagus. Probably, the thyrohyoid muscle, connecting the larynx to the hyoid apparatus, plays a major role in this active, rostrally directed movement of the larynx (Fig. 1.11).

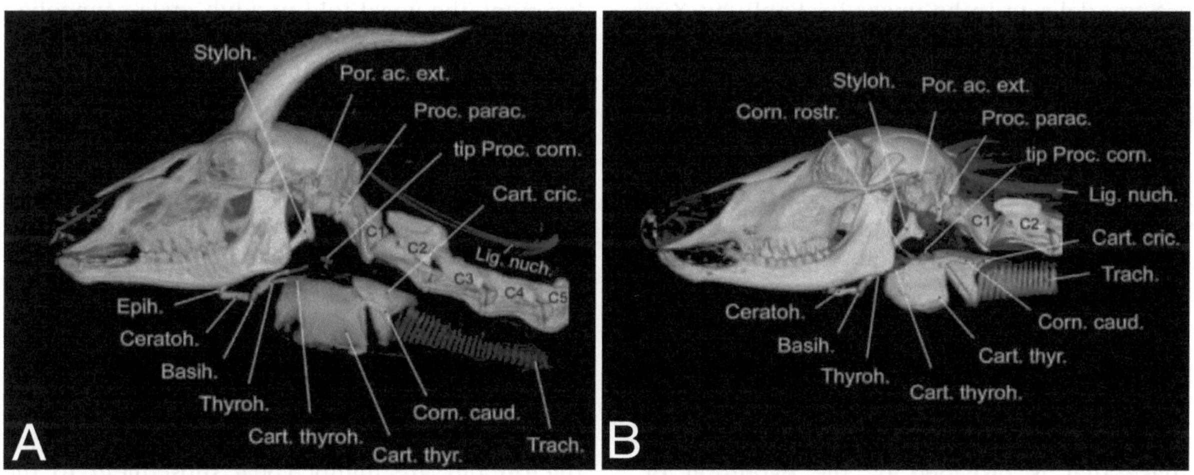

FIG. 1.3. Sexual dimorphism of the laryngeal cartilages as revealed by CT-based 3D imaging. Male (**A**) and female (**B**). (**A** alterated after Frey and Gebler 2003) (see Color Plates, Fig. 1.3AB).

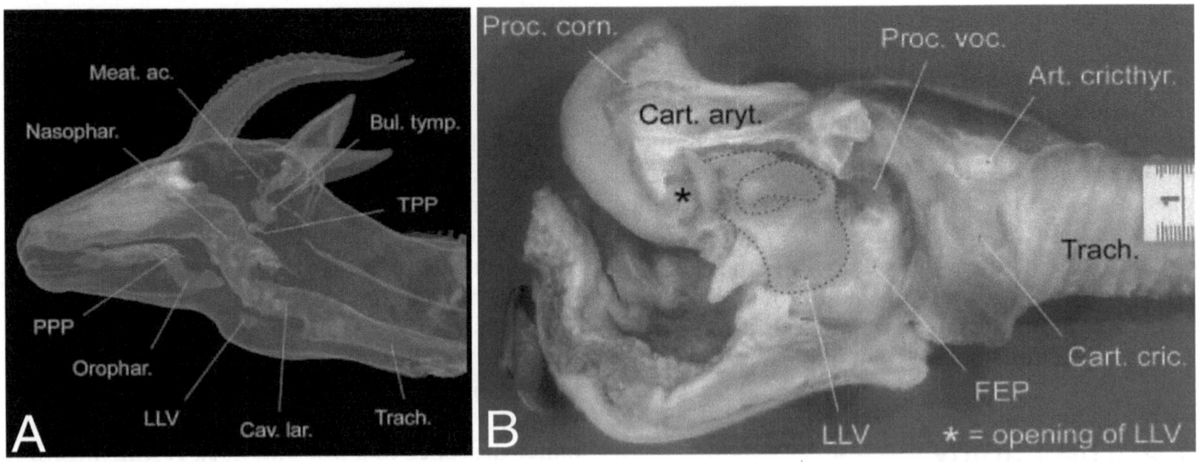

FIG. 1.4. Incomplete representation of lateral laryngeal ventricle (LLV) in a CT-based air mode 3D image, probably resulting from post-mortem ventricle collapse (**A**). Dissected left LLV of a male consisting of a crescent-shaped main chamber and a laterally protruding subsidiary sac (**B**). (**B** alterated after Frey and Riede 2003) (see Color Plates, Fig. 1.4B).

The broad (30–40 mm wide) keel of the thyroid cartilage in the adult male possibly relates to the evolution of a paired lateral laryngeal ventricle (= ventricle of Morgagni), unique among ruminants. On both sides it comprises a half moon-shaped main sac and a connected subsidiary sac that protrudes laterally between the ventricularis and the vocalis muscles. The dorsoventrally orientated slit-like entrance to the lateral laryngeal ventricle is situated lateral to the arytenoid cartilage (Fig. 1.4). The walls of the lateral laryngeal ventricle consist of thin soft tissue and are too delicate to be imaged reliably by X-ray-based computed tomographic scanning. Even a virtual 3D reconstruction of air-filled spaces did not provide satisfactory results, most probably because of partial postmortem ventricle collapse (Fig. 1.4).

Lateral laryngeal ventricles in non-ruminant mammals, e.g. in many primates (Starck and Schneider 1960), may develop into large air sacs which are no longer accomodated inside the laryngeal cavity but extend into the pectoral and axillary region. Comparison with these air sacs suggest that the lateral laryngeal ventricle of adult male Mongolian gazelle can also be inflated. This might explain the necessity of increased intralaryngeal space that is provided by the evolution of a broad keel of the thyroid cartilage. Intercalation of this keel provides more space between the two lateral walls of the thyroid cartilage and might, thus, allow for an inflation of both lateral laryngeal ventricles without narrowing the laryngeal cavity too much.

The female has a typical thyroid cartilage, V-shaped in transverse section and without a keel. A lateral laryngeal ventricle is missing.

The large, voluminous and tough vocal fold of adult male Mongolian gazelle bulges out caudally. A rostrally directed flexible portion is lacking. The vocal cleft (Rima glottidis) is situated between the caudal edges of the vocal folds. Instead of a vocal ligament, the vocal fold of adult male Mongolian gazelles is supported by a unique, cymbal-like fibroelastic pad consisting of tough and resilient connective tissue. The cymbal-like circular portion is about 40 mm in diameter. Dorsally, this fibroelastic pad is connected to the short vocal process of the arytenoid cartilage. Ventrally, it is anchored to the keel of the thyroid cartilage by a rostrally extending stem-like portion (Fig. 1.5).

The large and tough vocal fold in adult male Mongolian gazelles are thought to decrease the fundamental frequency and to increase the amplitude of their rutting barks.

The vocal fold of the female is smaller and lacks a fibroelastic pad. Females seldom vocalize during the rut but are highly vocal subsequent to the calving period when communicating with their young.

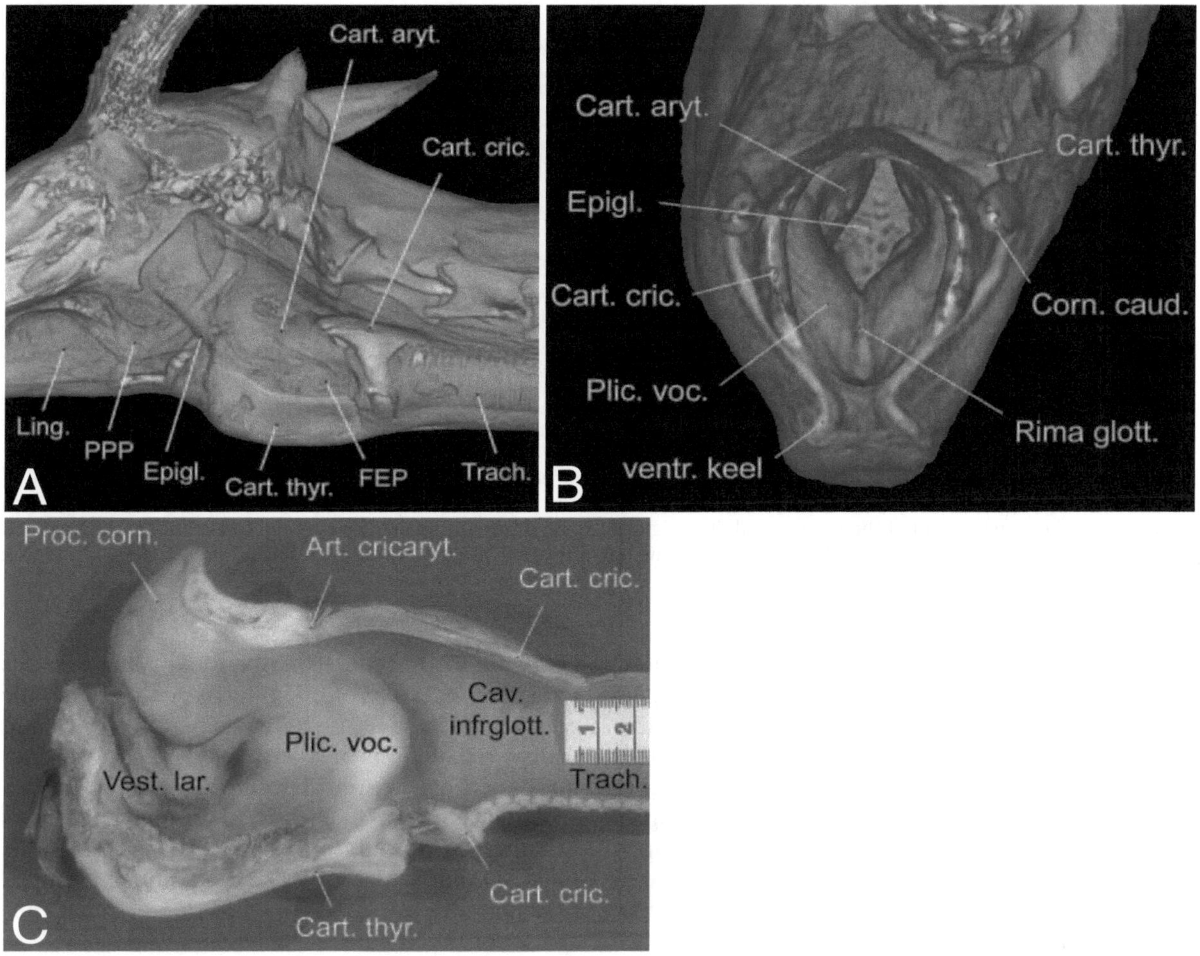

FIG. 1.5. Virtual parasagittal section of CT-based virtual 3D reconstruction of pharyngolaryngeal region of a male Mongolian gazelle (**A**) disclosing the connection of the arytenoid cartilage to the cymbal-shaped fibroelastic pad (FEP) supporting the vocal fold. Virtual transverse section through the larynx of A at the level of the vocal folds, caudal view (**B**). Mediosagittal section of larynx in medial view exposing the right vocal fold (**C**). (**C** alterated after Frey and Riede 2003) (see Color Plates, Fig. 1.5A–C).

1.3 The Pharynx

The pharynx of the adult male Mongolian gazelle has evolved two structures unique among ruminants: an unpaired palatinal pharyngeal pouch and a paired tympanic pharyngeal pouch.

The palatinal pharyngeal pouch consists of a ventromedian sac-like extension of the soft palate suspended between the root of the tongue and the epiglottis. Rostrally, it is fastened to the palatinal and pterygoidal bones. Classic anatomical dissection reveals that, in the resting state, the wall of the palatinal pharyngeal pouch is rather thick, shrunk and deeply wrinkled. Therefore, its contours become faintly visible in a 3D reconstruction based upon serial computed tomographic scans (Fig. 1.6).

The wrinkled surface of the palatinal pharyngeal pouch led Kleinschmidt (1961) to assume that it can be inflated. The palatinal pharyngeal pouch has a similar topographic position as the dulaa in male dromedarys (*Camelus ferus*, f. dromedarius) that is inflated and extruded from one side of the mouth in the course of rutting behaviour. However, there is no evidence so far, that the palatinal pharyngeal pouch of adult male Mongolian gazelles can be extruded through the mouth angle as the inflated dulaa of the dromedary

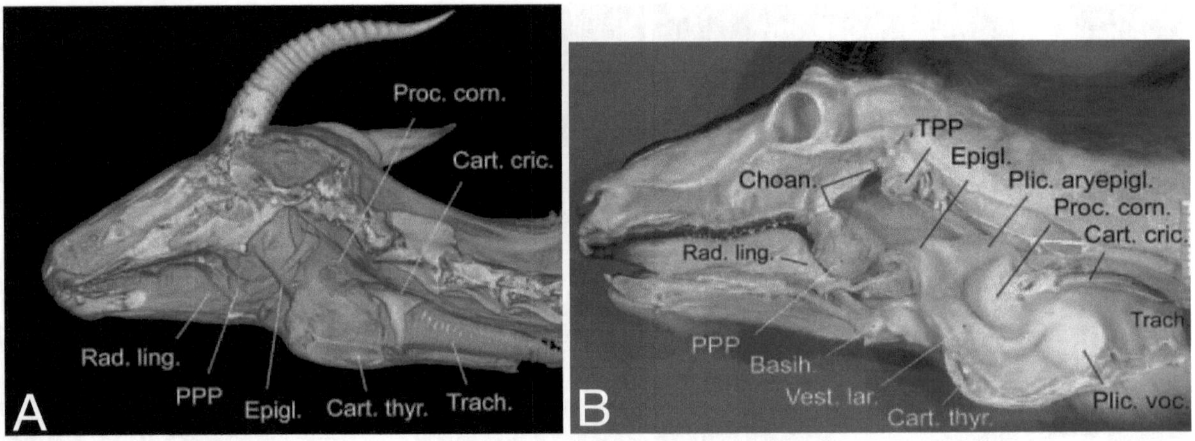

FIG. 1.6. Virtual parasagittal section of CT-based 3D reconstruction of the head of a male Mongolian gazelle (**A**) demonstrating presence and topographic position of the unpaired palatinal pharyngeal pouch (PPP). Dissection confirms its position between the root of the tongue and the epiglottis. In addition, the strongly wrinkled wall of the PPP is revealed (**B**) implying that some sort of inflation might occur. (**B** alterated after Frey and Gebler 2003) (see Color Plates, Fig. 1.6AB).

(Kleinschmidt 1962). We assume that an intra-pharyngeal inflation of the palatinal pharyngeal pouch is achieved in the adult male Mongolian gazelle. Evidence comes from video clips that show the occurrence of a momentary bulging of the ventral neck region rostral to the laryngeal prominence. This bulging might have escaped observation so far owing to its short-time appearance. Bulging of the neck occurs caudally adjacent to the angle of the lower jaw.

Unfortunately, a convincing functional mechanism of the intra-pharyngeal inflation of the palatinal pharyngeal pouch cannot be provided so far. Apart from that, it is unclear whether the extremely short inflation of the palatinal pharyngeal pouch serves to produce or to accompany any acoustic output.

In addition to the palatinal pharyngeal pouch, adult male Mongolian gazelles have evolved a paired tympanic pharyngeal pouch. It is thin-walled and inflatable and consists of a sac-like lateral protrusion of the dorsal pharyngeal wall between the base of the stylohyoideum and the tympanic bulla of the skull. It might represent an outpouching of the auditary tube connecting the tympanic cavity with the nasopharynx (Fig. 1.7). After artificial inflation to an extent that it tightly fits the space provided by surrounding structures, the tympanic pharyngeal pouch measured 25 mm both in dorsoventral and rostrocaudal width.

Owing to its delicate wall and adjacent osseous structures (stylohyoid, skull base) the tympanic pharyngeal pouch is only visible in computed transverse sections of X-ray based imaging but not in 3D reconstructions. Therefore, it has to be dissected according to classic anatomical methods and imaged directly in order to get a better idea of its 3D shape (Fig. 1.7). The tympanic pharyngeal pouches might serve as additional resonance spaces of the vocal tract.

The female has a primitive pharynx lacking any additional side spaces.

Apart from PPP and TPP, the pharynx of the adult male Mongolian gazelles is characterised by its remarkable resilience that, presumably, has evolved in connection with the intermittent retraction of the larynx simultaneous to the emission of loud head-up rutting barks. Movement of the larynx down to the thoracic aperture requires a high flexibility of the pharyngeal walls to allow for a concomitant pronounced elongation of the pharynx. Resilience of the pharynx might assist in the return of the larynx to its resting position. Descent of the larynx to its low resting position during the rut will increase the space available for the PPP and might, therefore, be advantageous for its above mentioned intra-pharyngeal inflation. Pronounced retraction of the larynx does not occur in the female and, thus, the pharynx of the female is not subjected to comparable straining forces as that of the male.

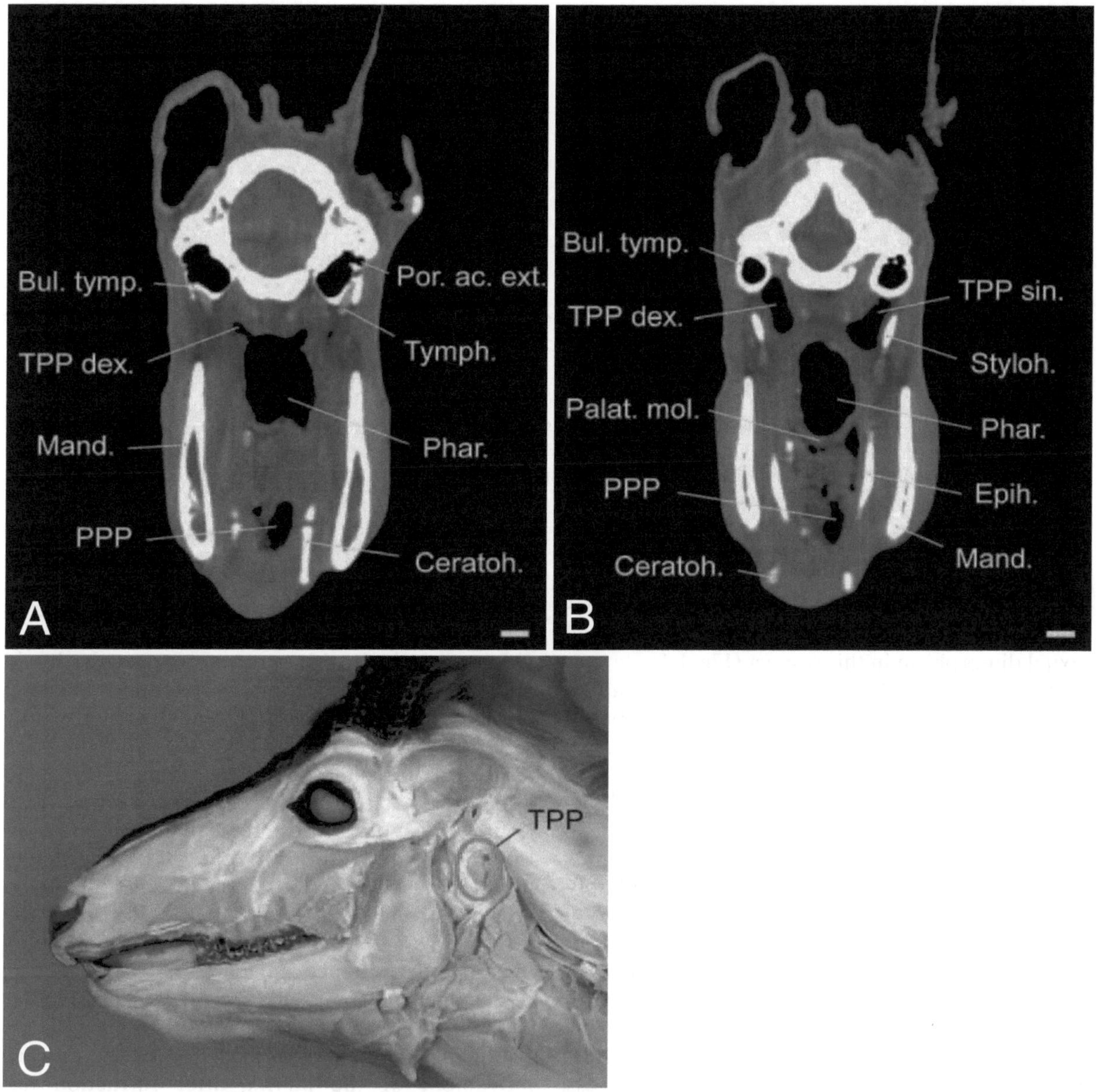

FIG. 1.7. CT transverse sections of the head of a male Mongolian gazelle showing origin of paired tympanic pharyngeal pouch (TPP) from the roof of the pharynx (**A**) and contact of TPP with tympanic bulla (**B**). Dissected topographic position of TPP (**C**) between tympanohyoid (rostrally), stylohyoid (laterally) and origin of occipitohyoid muscle (caudally) (see Color Plates, Fig. 1.7C).

1.4 The Hyoid Apparatus

The structure of the hyoid apparatus is in accordance with the basic ruminant plan. However, the thyrohyoid is elongated caudally by a rod-like cartilaginous element. The caudal end of this weak and flexible thyrohyoid cartilage connects to the rostral horn of the thyroid cartilage to form the thyrohyoid articulation. This junction is remarkably loose so that a high flexibility can be inferred. During the rut the thyrohyoid articulation apparently gets transformed into a ligament-like structure temporarily and restores itself to a less flexible state thereafter. Evidence for a ligamentous thyrohyoid

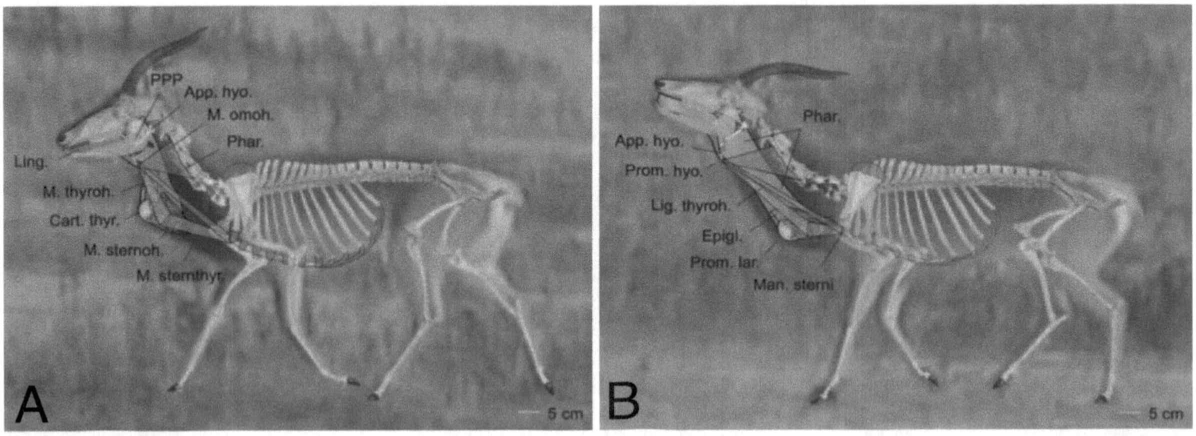

FIG. 1.8. Low resting position of the larynx during the rut (**A**) and maximally retracted larynx during a head-up bark of a male Mongolian gazelle (**B**). In this stage the hyoid apparatus is tilted caudally and ligament-like thyrohyoid connection and pharynx are maximally extended. Skeleton and respective organs were fitted to video single frames (see Color Plates, Fig. 1.8AB).

connection in adult male Mongolian gazelles comes from Pallas (1777) who points to a pronounced sexual dimorphism in this respect (Fig. 1.1).

Owing to its ossified elements, the hyoid apparatus is easily visible in 3D reconstructions based upon computed tomographic serial scans (Fig. 1.3) and also in respective virtual sections.

The hyoid apparatus must be capable of following the pronounced retraction of the larynx in adult male Mongolian gazelles during the emission of their loud head-up rutting barks. Probably, this is achieved by backward tilting and extension involving differential contraction and extension of the paired sternohyoid and thyrohyoid muscles. The tilting is made possible by the resilience of the non-ossified tympanohyoid. At the same time it is limited owing to the caudally adjacent occipitohyoid muscle and the paracondylar process. Pronounced extension is made possible by cartilaginous connections between the remaining hyoid elements.

The extension of the complete hyoid apparatus will cause a synchronous lowering of the tongue which is muscularly attached to cerato- and basihyoid. Therefore, the torus of the tongue, touching the hard palate in its resting position, will be displaced ventrally, thereby opening a wider gap between itself and the hard palate.

Kleinschmidt (1961) noted a hyoid prominence in the male specimen that he examined. This observation corresponds with the observation of a small neck expansion during maximal larynx retraction in the live animal, discovered by video analysis. In this stage we expect the hyoid apparatus to be maximally extended and the basihyoid to push against the neck skin rostrally (Fig. 1.8).

Return of the larynx and re-folding of the hyoid apparatus back to their respective resting positions will require differential extension and contraction of the sternothyroid and sternohyoid muscles plus contractions of the thyrohyoid muscle and the strong ceratohyoid muscle. Return of the larynx is assisted by re-shortening of the pharynx owing to its resilience and, perhaps, by contraction of the pharyngeal musculature (cf. Frey and Gebler 2003).

Possibly, a considerable portion of the upward movement of the larynx, after retraction has ended, is not achieved by active muscular contraction but by the resilience of soft tissues that before had become extended during retraction, e.g. pharynx and tongue fixations. Additional passive protraction may come from resilient re-folding tendencies of the extended hyoid cartilages. The trachea will be protracted passively by its connection to the larynx.

Lack of laryngeal mobility in the female makes specific adaptations of the hyoid apparatus unnecessary. The thyrohyoid cartilage is shorter than in the male and, in contrast to the male, the thyrohyoid articulation is constantly short.

1.5 Ventral Hyoid and Extrinsic Laryngeal Muscles

Except for some tendinous structures, muscles cannot be accurately detected by X-ray-based imaging. Accordingly, they have to be dissected by classic anatomical methods and figured by direct photographic documentation.

The superficial parotidoauricularis muscle is part of the cutaneous muscles of the head. It originates from the parotidomasseteric fascia and, ventromedially, from the fibres of its contralateral counterpart. Its fibres continuously converge in dorsal direction towards its insertion to the auricular concha. The connected broad ventral portions of both parotidoauricularis muscles support the epiglottal region of the larynx in a hammock-like sling (Fig. 1.9).

In contrast to other ruminants, the powerful mandibular portion of the sternocephalic muscle is double-bellied. Both portions originate from the most rostral portion of the sternum. The stronger ventral portion inserts with a strong tendon to the superficial aponeurosis of the masseter muscle and to the superficial connective tissue covering the buccal region and the mandibula. The weaker dorsal portion inserts via an aponeurosis, which is traversed be the external jugular vein, to the parotid fascia and, superficially, to the ventral edge of the cleidocephalic muscle at the level of the transverse process of the atlas (Fig. 1.10).

The broad and strong, triangular omohyoid muscle arises from fasciae lateral to the transverse processes of cervicals C_2–C_4. It passes deep to the external jugular vein and, together with the sternohyoid muscle, inserts to the basihyoideum (Fig. 1.10).

Sandwiched between the ventral portion of the sternocephalic muscle laterally and the sternothyroid and thyrohyoid muscles medially, the slender sternohyoid muscle passes as a long, narrow band along the trachea and the keel of the thyroid cartilage and inserts to the basihyoideum, ventral to the insertion of the omohyoid muscle (Fig. 1.10).

The sternothyroid muscle, also a long band but broader and stronger than the sternohyoid muscle, originates from the most rostral end of the sternum, medial to the origin of the sternocephalic muscle. It passes laterally along the trachea touching the ventral contour of the thyroid gland. The sternothyroid muscle inserts onto the caudal edge and the caudolateral surface of the thyroid cartilage. Its insertion is separated from the origin of the thyrohyoid muscle merely by a small tendinous strip (Fig. 1.11).

The thyrohyoid muscle arises immediately rostral to the insertion of the sternothyroid muscle from the lateral surface of the thyroid cartilage. It broadens and covers most of the lateral surface of the enlarged thyroid cartilage and the insertion of the caudal constrictor muscles of the pharynx.

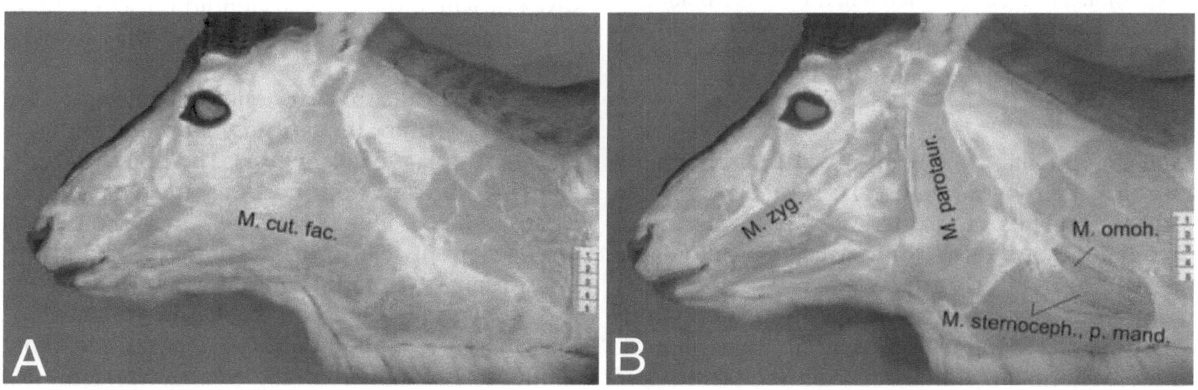

FIG. 1.9. Delicate cutaneous muscle of the head of a male Mongolian gazelle (**A**). Parotidoauricularis muscle, omohyoid muscle and the two portions of the strongly developed sternocephalic muscle after removal of the cutaneous muscle of the head (**B**). The parotidoauricularis muscle supports the epiglottal region in a hammock-like sling. Omohyoid and mandibular portion of sternocephalic muscle lend lateral support to the large and heavy larynx (see Color Plates, Fig. 1.9AB).

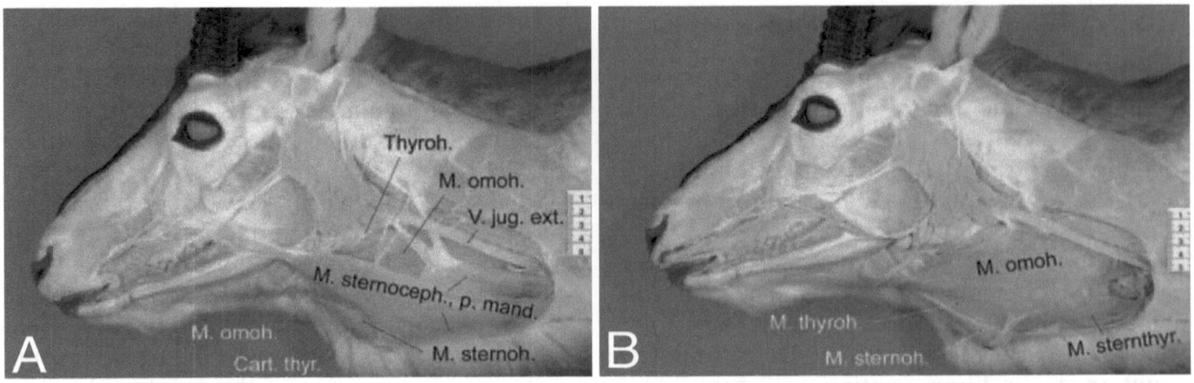

FIG. 1.10. The double-bellied mandibular portion of the sternocephalic muscle has an additional aponeurotic insertion, traversed by the external jugular vein, to the parotid fascia and to the ventral edge of the neck muscles (**A**). After removal of the mandibular portion of the sternocephalic muscle, the broad omohyoid muscle is exposed (**B**). It covers the pharyngolaryngeal region laterally. Ventrally, the omohyoid muscle is connected with its contralateral counterpart by connective tissue (see Color Plates, Fig. 1.10AB).

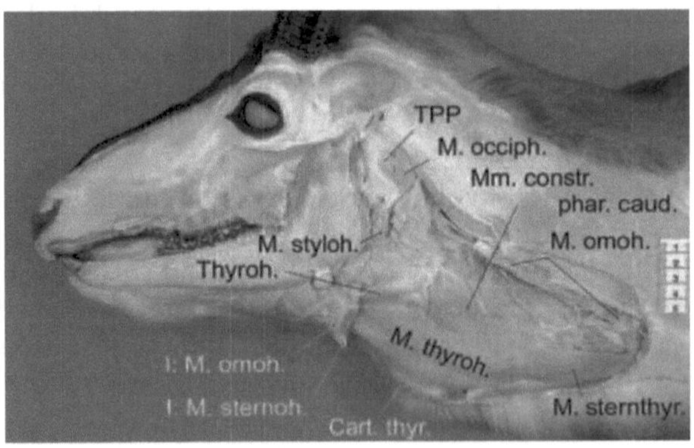

FIG. 1.11. Deep musculature of the laryngeal region of a male Mongolian gazelle after removal of the omohyoid muscle. The larynx is connected to the sternum by the sternothyroid muscle and to the hyoid apparatus by the thyrohyoid muscle (see Color Plates, Fig. 1.11).

The exceptional dorsoventral width and strength of the thyrohyoid muscle is not found in other ruminants. It inserts not only to the thyrohyoideum but also onto the basihyoideum (Fig. 1.11).

In correspondence with the enlarged epiglottis of the male Mongolian gazelle, its hyoepiglottic muscle is also large. It arises from the cerato- and basihyoideum and inserts onto the rostroventral surface of the epiglottis.

Corresponding muscles in the female are much smaller and weaker than in the male and not double-bellied.

The strong muscles, adjacent to the male larynx, not only support the enlarged larynx laterally and enhance its suspension from the skull base and the neck. Apart from that, video clips, made during the rut in the natural habitat, document that male Mongolian gazelles, simultaneous to producing their loud head-up rutting barks, pronouncedly retract their larynges (Fig. 1.8).

1.6 Intrinsic Laryngeal Muscles

There are five intrinsic laryngeal muscles in both sexes of which the thyroarytenoid muscle is sexually dimorphic. We shall focus on this sexually dimorphic muscle.

In the male, the thyroarytenoid muscle is divided into two portions: the rostral ventricularis muscle and the caudal vocalis muscle. The dorsoventrally coursing fibres of both portions are separated by a pronounced cleft through which the lateral laryngeal ventricle protrudes laterally. Both portions originate from the laterodorsal surface of the thyroid cartilage's broad keel, the ventricularis muscle more rostrally, the vocalis muscle more caudally. The fibres of the ventricularis muscle insert onto the lateral surface of the arytenoid cartilage including the prominent arycorniculate ligament, up to the arcuate crest and rostral to the muscular process. The fibres of the vocalis muscle insert onto the lateral surface of the arytenoid

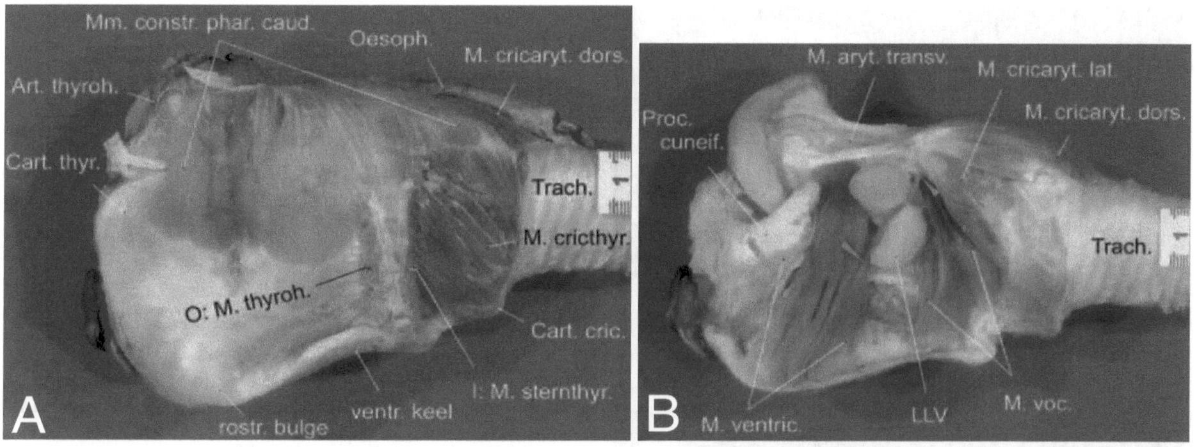

FIG. 1.12. The intrinsic laryngeal muscles of a male Mongolian gazelle: the most superficial one is the cricothyroid muscle (**A**). After removal of the left half of the thyroid cartilage, the transverse arytenoid muscle, the dorsal and lateral cricoarytenoid muscles and the thyroarytenoid muscle are exposed. The latter is divided into a rostral ventricularis and a caudal vocalis muscle. Between them the two-chambered inflatable LLV projects laterally (**B**). (**B** alterated after Frey and Gebler 2003 and Frey and Riede 2003) (see Color Plates, Fig. 1.12AB)

cartilage, ventral to the base of its muscular process. The bundles caudal to the connection between vocal process and fibroelastic pad are particularly strong (Fig. 1.12).

In the female, conforming to the primitive situation in ruminants, the thyroarytenoid muscle forms an almost uniform muscular plate extending between the un-keeled thyroid cartilage and the arytenoid cartilage.

1.7 Vocal Tract Length (VTL)

So far, a pronounced retraction of the larynx during rutting roars of adult males has only been reported for some cervid species, e.g. red and fallow deer (*Cervus elaphus*, *Dama dama* – Fitch and Reby 2001; Reby et al. 2005; McElligott et al. 2006). The same authors demonstrated that larynx retraction causes a vocal tract elongation and a simultaneous decrease of vocal tract resonances (formants). This acoustic effect is thought to signal larger body size and better quality, e.g. fighting abilities, to male and female conspecifics.

The polygynous, harem-like mating system of Mongolian gazelle, where one dominant male rounds up several females, is well comparable with the mating system of red and fallow deer (cf. Clutton-Brock and Albon 1979). Therefore, a similar function of larynx retraction in Mongolian gazelle is assumed.

Apart from larynx retraction, the lowered resting position of the larynx in male Mongolian gazelle for the duration of the rut, also influences resonance characteristics of the vocal tract. Accordingly, a comparison of the summer barks of adult males with loud head-up barks during the rut in winter demonstrate a lowering of corresponding vocal tract resonances (Frey and Gebler 2003; Frey et al. 2008 in preparation).

VTL can be measured either virtually in sagittal sections of computed tomographic 3D reconstructions or directly in classical dissections. Generally, virtual measurements render smaller values owing to preservation of original topographic positions. In a real dissection, half of the specimen, e.g. the left halves of the hyoid apparatus and larynx, have to be removed before VTL (from lips to vocal folds) can be measured. As a consequence, connections between structures to be measured loosen and distances become larger. VTL in the male Mongolian gazelle can also be estimated by single frame analysis of video clips owing to the pronounced laryngeal prominence produced by the evolutionarily enlarged larynx. In the male sex, combined results suggest a resting state VTL of about 300 mm in summer and about 400 mm in winter. In the female, the VTL constantly amounts to about 250 mm (Fig. 1.13).

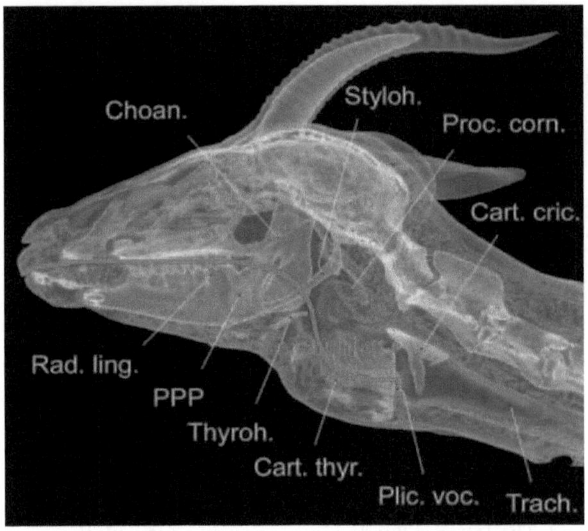

Fig. 1.13. Measurement of vocal tract length (VTL = lips to glottis *in red*) in a virtual mediosagittal section of a computed 3D reconstruction of the head of a male Mongolian gazelle. Owing to the formation of a fibroelastic pad, the vocal fold can be clearly identified between the caudal edge of the thyroid cartilage and the rostral edge of the cricoid cartilage. (alterated after Frey and Gebler 2003) (see Color Plates, Fig. 1.13).

1.8 Evolution

Polygynous, harem-like mating systems favour the evolution of pronounced sexual dimorphisms by the effects of sexual selection (Andersson 1994). Probably, this also applies to the unique vocal anatomy of the male Mongolian gazelle. Initially, a mutation might have led to a slightly larger and/or mobile larynx in one male. Accordingly, this male was able to acoustically pretend a larger body size (better fighting ability) to male and female conspecifics (cf. Fitch and Reby 2001) resulting in a competitive advantage over rival males and a higher attractiveness towards potential female mating partners. Ultimately, this led to an immediate increase of its reproductive success.

Once this evolutionary process had been started, males could not 'afford' an ordinary-sized larynx, as acoustically less impressive vocalizations would have directly reduced their reproductive success. This, via male-male competition (cf. Reby et al. 2005) and female choice (cf. Charlton et al. 2007), may have resulted in an evolutionary arms race (Dawkins and Krebs 1979) among males that did not end before enlargement and specializations of the male larynx and vocal tract were counterselected by natural selection. Owing to this evolutionary constraint that limited further elaboration of the vocal apparatus and that holds for all males equally, their rutting calls remain honest indications of body size, although on a different acoustic level than before onset of this evolutionary process.

References

Andersson M (1994) Sexual selection. Monographs in behavior and ecology. Princeton University Press, Princeton, NJ

Clutton-Brock TH, Albon SD (1979) The roaring of red deer and the evolution of honest advertising. Behaviour 69:145–170

Charlton BD, Reby D, McComb K (2007) Female red deer prefer the roars of larger males. Biol Lett 3:382–385, doi: 10.1098/rsbl.2007.0244

Dawkins R, Krebs JR (1979) Arms races between and within species. Proc R Soc Lond B 205:489–511

Fitch WT, Reby D (2001) The descended larynx is not uniquely human. Proc R Soc Lond B 268:1669–1675

Frey R, Gebler A (2003) The highly specialized vocal tract of the male Mongolian gazelle (*Procapra gutturosa* Pallas, 1777 – Mammalia, Bovidae). J Anat 203:451–471

Frey R, Riede T (2003) Sexual dimorphism of the larynx of the Mongolian gazelle (*Procapra gutturosa* Pallas, 1777) (Mammalia, Artiodactyla, Bovidae). Zool Anz 242:33–62

Frey R, Gebler A, Olson KA, Odonkhuu D, Fritsch G, Batsaikhan N, Stuermer IW (2008) Mobile larynx in Mongolian gazelle. Retraction of the larynx during rutting barks in male Mongolian gazelle (*Procapra gutturosa* Pallas, 1777). (in preparation)

Kleinschmidt A (1961) Über die große mongolische Kropfgazelle (*Procapra gutturosa* Pallas 1777). Stuttgarter Beitr Naturk 79:1–24

Kleinschmidt A (1962) Die Schlund- und Maulblase (Bursa faucium) bei Kamel und Großer Kropfgazelle. Die Naturwissenschaften 49:140–141

McElligot AG, Birrer M, Vannoni E (2006) The retraction of the mobile descended larynx during groaning enables fallow bucks (*Dama dama*) to lower their formant frequencies. J Zool (London) 270:340–345

Pallas PS (1777) Ad Genus Antiloparum complementum, 6–21. *Antilope gutturosa*, 46–51, Descriptio *Antilopes gutturosae*, 52–60, Taf. II et III (fig. 13–17). Spicilegia Zoologica quibus novae imprimis et obscurae Animalium Species Iconibus, Descriptionibus atque Commentariis illustrantur **12**, Christianum Fridericum Voss, Berlin

Reby D, McComb K, Cargnelutti B, Darwin C, Fitch WT, Clutton-Brock T (2005) Red deer stags use formants as assessment cues during intrasexual agonistic interactions. Proc R Soc 272:941–947

Starck D, Schneider R (1960) Respirationsorgane, A. Larynx. *Primatologia* 3/2:423–587

Abbreviations

App. hyo.	=	hyoid apparatus
Art. cricaryt.	=	cricoarytenoid articulation
Art. cricthyr.	=	cricothyroid articulation
Art. thyroh.	=	thyrohyoid articulation
Basih.	=	basihyoid
Bul. tymp.	=	tympanic bulla
C1–C5	=	cervical vertebrae 1–5
Cart. aryt.	=	arytenoid cartilage
Cart. cric.	=	cricoid cartilage
Cart. thyr.	=	thyroid cartilage
Cart. thyroh.	=	thyrohyoid cartilage
Cav. infrglott.	=	infraglottic cavity
Cav. lar.	=	laryngeal cavity
Ceratoh.	=	ceratohyoid
Choan.	=	choanae
Corn. caud.	=	caudal horn
Corn. rostr.	=	rostral horn
Epigl.	=	epiglottis
Epih.	=	epihyoid
FEP	=	fibroelastic pad
I: M. omoh.	=	insertion of omohyoid muscle
I: M. sternoh.	=	insertion of sternohyoid muscle
I: M. sternthyr.	=	insertion of sternothyroid muscle
Lig. nuch.	=	nuchal ligament
Lig. thyroh.	=	thyrohyoid ligament
Ling.	=	tongue
LLV	=	lateral laryngeal ventricle
Man. sterni	=	sternal manubrium
Mand.	=	mandibula (dentary)
Meat. ac.	=	acoustic meatus
M. aryt. transversus	=	transverse arytenoid muscle
M. cricaryt. dors.	=	dorsal cricoarytenoid muscle
M. cricaryt. lat.	=	lateral cricoarytenoid muscle
M. cricthyr.	=	cricothyroid muscle
M. cut. fac.	=	facial cutaneous muscle
M. occiph.	=	occipitohyoid muscle
M. omoh.	=	omohyoid muscle
M. parotaur.	=	parotidoauricularis muscle
M. sternoceph., p. mand.	=	mandibular portion of sternocephalic muscle
M. sternoh.	=	sternohyoid muscle
M. sternthyr.	=	sternothyroid muscle
M. styloh.	=	stylohyoid muscle
M. thyroh.	=	thyrohyoid muscle
M. ventric.	=	ventricularis muscle
M. voc.	=	vocalis muscle
M. zyg.	=	zygomatic muscle
Mm. constr. phar. caud.	=	caudal constrictor muscles of pharynx
Nasophar.	=	nasal portion of pharynx
O: M. thyroh.	=	origin of thyrohyoid muscle
Oesoph.	=	oesophagus
Orophar.	=	oral portion of pharynx
Palat. mol.	=	soft palate
Phar.	=	pharynx
Plic. aryepigl.	=	aryepiglottic fold
Plic. voc.	=	vocal fold
Por. ac. ext.	=	external acoustic meatus
PPP	=	palatinal pharyngeal pouch
Proc. corn.	=	corniculate process
Proc. cuneif.	=	cuneiform process of epiglottis
Proc. parac.	=	paracondylar process
Proc. voc.	=	vocal process
Prom. hyo.	=	hyoid prominence
Prom. lar.	=	laryngeal prominence
Rad. ling.	=	root of the tongue
Rima glott.	=	glottic cleft
rostr. bulge	=	rostral bulge of thyroid cartilage
Styloh.	=	stylohyoid
Thyroh.	=	thyrohyoid
tip Proc. corn.	=	tip of corniculate process
Tymph.	=	tympanohyoid
TPP (dex., sin.)	=	(right, left) tympanic pharyngeal pouch
Trach.	=	trachea
V. jug. ext.	=	external jugular vein
ventr. keel	=	ventral keel of thyroid cartilage
Vest. lar.	=	laryngeal vestibulum

2 Anatomical Peculiarities of the Vocal Tract in Felids

Gerald E. Weissengruber[1], Gerhard Forstenpointner[1], Sandra Petzhold[1], Claudia Zacha[1], and Sibylle Kneissl[2]

2.1 Introduction

The fourty species (Wilson and Reeder 2005) of felids (family *Felidae*) are all strictly carnivorous and perfectly designed to capture live prey. The family is divided into two subfamilies, i.e., the *Felinae* and the *Pantherinae* (Wilson and Reeder 2005). The majority of adult felids live and hunt separately (Sunquist and Sunquist 2002). Besides other morphological adaptations to hunting (e.g., structure of the vertebral column or the limbs) cats possess a foreshortened face, powerful jaw muscles, strong canines and secodont molars but the number of teeth is less than in other carnivores like dogs or bears (Sunquist and Sunquist 2002). Felids vary enormously in size. Large size differences within a particular species are found in species with broad geographic distributions (Sunquist and Sunquist 2002). Although most of the cat species are solitary, felids use visual signals (e.g., cheek rubbing), odours (secretions from various glands, saliva, urine, faeces) and a variety of vocalizations for communication within a certain social system (Sunquist and Sunquist 2002).

The majority of vocalizations is most likely generated by oscillations of the vocal folds and some structural characters of the calls may be influenced by the structure and length of the so-called upper vocal tract, i.e., the length of the cavities between lips and glottis (Weissengruber et al. 2002). Cats use a set of close-range, medium-range and long-distance calls (Peters and Tonkin-Leyhausen 1999). The repertoire of vocalizations in felids includes discrete signal types (e.g., hiss) and graded systems (e.g., mew/main call). Purring, a close-range vocalization which is well known in the domestic cat (*Felis catus*), has been reported also in *Viverridae* but within the family *Felidae* it is restricted to the *Felinae* (Peters 1981; 2002; Sunquist and Sunquist 2002). On the other hand, the impressive roaring sequence similar to that of the lion occurs only in the jaguar and leopard (Sunquist and Sunquist 2002). Following the definitions of G. Peters, the roaring sequence is composed of two vocalization types, main calls with grunt element and grunts. The main call with grunt element is a predominantly intense vocalization which is present only in four species of the *Pantherinae* namely the lion, leopard, jaguar and tiger (Sunquist and Sunquist 2002). It has been hypothesized that the presence or absence of roaring and purring in the vocal repertoires of felids is correlated with differences in the structure of the hyoid apparatus (Peters and Hast 1994). Comparative anatomist Richard Owen was the first who mentioned in 1833 an elastic ligament being part of the hyoid suspensory apparatus in the lion and that the subsequent far caudal position of the larynx might relate to the 'roar of the lion' (Owen 1835). Pocock (1917) included all felids in which the 'suspensorium of the hyoid' is 'imperfectly ossified', i.e., lion (*Panthera leo*), tiger (*Panthera tigris*), leopard (*Panthera pardus*), jaguar (*Panthera onca*) and snow leopard (*Uncia uncia*) in the subfamily *Pantherinae*. In these species, the bony epihyoid is replaced by an elastic ligament. For several decades these five felid species have been designated as the 'roaring cats', although the vocalization type 'roaring' has not been defined by means of using bioacoustical methods. Following a recent

[1]Institute of Anatomy, [2]Clinic for Diagnostic Imaging, University of Veterinary Medicine Vienna, Veterinärplatz 1, 1210 Wien, Austria

taxonomic reference the clouded leopard (*Neofelis nebulosa*) is also a member of the *Pantherinae*, although its suspensory apparatus of the hyoid is 'completely ossified' as that of the *Felinae* (Wilson and Reeder 2005).

The low (caudal) position of the larynx in humans, red and fallow deer and in *Pantherinae* is hypothesized to extend the upper vocal tract and to lower the formant frequencies of vocalizations (Fitch and Reby 2001; Weissengruber et al. 2002; McElligott et al. 2006). In mammals, lowered (decreased) formant frequencies serve most likely to exaggerate perceived body size (Fitch and Reby 2001; McElligott et al. 2006). Furthermore, lower frequency sounds travel further than sounds with higher frequencies and thus the evolution of descended larynges might have resulted from the need of more effective sound propagation over longer distances (McElligott et al. 2006). As the sound signal is produced in the larynx and subsequently filtered by the vocal tract, it is obvious that the ossification status of the suspensory apparatus of the hyoid does not necessarily cause a species' ability to roar or to purr (Peters and Hast 1994; Peters 2002). Containing a large 'fibroelastic' pad-like structure, the long and voluminous vocal folds of *Pantherinae* cause most likely the low fundamental frequency of distinct vocalizations such as the main call with grunt element of the roaring sequence (Peters and Hast 1994; Weissengruber et al. 2002). Purring is caused by rapid twitching of the vocalis muscle, whereas the large pads within the vocal folds of *Pantherinae* might impede rapid contractions of this muscle and thus make it difficult to purr (Frazer Sissom et al. 1991; Weissengruber et al. 2002).

In this chapter we focus on structures of the larynx, the pharynx, the hyoid apparatus and their topographical positions in certain felid species (lion, tiger, leopard, jaguar, snow leopard, clouded leopard, cheetah, Asiatic golden cat, jungle cat, black footed cat, domestic cat, wildcat, pampas cat, Geoffroy's cat, oncilla, serval, Eurasian lynx, leopard cat, puma, jaguarundi). The specimens available (most of them from the Museum Alexander Koenig, Bonn, Germany) were investigated using dissection, radiographs, CT- and MR-imaging and histological techniques. Combining classical methods such as dissection and histological processing with modern imaging techniques serves as a sound background for interpreting acoustic data.

2.2 Larynx

The larynx contains the oscillating elements responsible for sound generation. Accordingly, the structure of the larynx plays an important role in vocal communication. As in other mammals, the skeleton of the felid larynx consists of the thyroid, cricoid, arytenoid and epiglottic cartilages. Size and shape of the cartilages and the formation of intercartilaginous as well as connective structures with the thyrohyoid (e.g., joints, ligaments) vary between species. Apart from the mentioned cartilages, the larynx comprises several further types of tissues: muscles, ligaments, membranes, glands, tonsils, lymph nodules and a mucosal lining. The shape and structure of the laryngeal cavity walls have an impact on sound production and the bioacoustic characters of vocalization. In felids, the lateral ventriculus laryngis between the vestibular and the vocal fold is either missing or comprises a rather shallow groove. In all cat species the vestibular folds are rather thin and can be sharp-edged. The vocal folds or parts of it in species of the subfamily *Pantherinae* are extremely voluminous and widened. This enlargement of the vocal folds is mainly caused by the occurrence of a large pad-like structure underneath the mucosa (Fig. 2.1) consisting of collagenous and elastic fibres

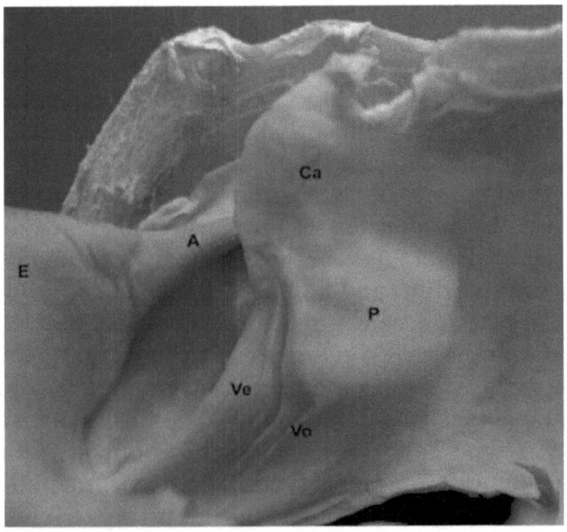

FIG. 2.1. Larynx of the tiger. Medial view. *E*, epiglottis; *A*, Plica aryepiglottica; *Ca*, Cartilago arytenoidea (covered by mucosa); *P*, pad; *Ve*, Plica vestibularis; *Vo*, Plica vocalis (see Color Plates, Fig. 2.1).

and a large amount of glycosamino-glycans. Glycosamino-glycaus contribute to the maintenance of an optimal tissue viscosity and stiffness that is important in oscillating systems. The pads are located within the dorsal part of the vocal folds close to the ventral border of the arytenoid cartilages. Contrary to Hast (1986), who stated that in the larynx of the snow leopard a pad of fibroelastic tissue is missing, the occurrence of a pad could be verified in two of three snow leopards examined in the present study. This difference in findings is possibly due to the fact that Hast (1986, 1989) investigated the rostral border of the Plica vestibularis ('rostral portion of undivided vocal folds') and not the Plica vocalis itself. Even in the vocal folds of clouded leopards (*Neofelis nebulosa*) a pad-like structure, similar in size to that of other *Pantherinae*, is macroscopically visible. Although the hyoid of the clouded leopard is 'completely ossified' (as in any species of the subfamily *Felinae*), it has been assigned to the *Pantherinae* (Wilson and Reeder 2005). Relatively smaller pad-like structures within the vocal folds can be found in many species of the subfamily *Felinae*, such as the jungle cat (*Felis chaus*), the pampas cat (*Leopardus colocolo*), the black footed cat (*Felis nigripes*) or the serval (*Leptailurus serval*). In almost all examined specimens of *Felinae*, the thinner Plica vocalis (compared to *Pantherinae*) is separated by an obvious Ventriculus laryngis from the often sharp-edged Plica vestibularis (e.g., in the black footed cat).

Increase of the oscillating mass or of the length of vocal folds leads to lower fundamental frequencies and thus to a lower pitch of vocalizations. Vocal fold length and/or mass is not only related to structural peculiarities but it may also be influenced by the body size of the species. The main call with grunt element of the tiger, which is the largest felid species and possesses a large pad within its vocal folds, reveals a very low fundamental frequency. On the other hand, the main call of one of the smallest cat species, the black footed cat (*Felis nigripes*, body weight up to 2.4 kg), is also low pitched which might be related to the occurrence of a pad within its vocal folds. Besides the occurrence, shape and stiffness of pads within vocal folds, occurrence and tension of fibres or vocal ligaments and shape and tension of the thyroarytenoid muscle have an effect on the oscillating properties of vocal folds. Furthermore, central nervous control and neuromuscular regulation of the activities of laryngeal, oropharyngeal and respiratory muscles represent crucial points in vocal communication (Shiba et al. 1997a,b). A comprehensive interpretation of neurophysiological mechanisms and effects on sound production caused by tissue properties is still missing.

2.3 Pharynx

The pharynx is the junction of the respiratory and digestive tracts. It is located between the oral and nasal cavities rostrally and the esophagus and larynx caudally. In the domestic cat and the cheetah, representatives of the subfamily *Felinae*, the Cavum pharyngis shows no anatomical peculiarities in comparison with other domestic mammals. In adult specimens of the lion, the tiger, the leopard and the jaguar (other species of the *Pantherinae* have not yet been examined in this respect) the pharyngeal cavity and its musculomembranous walls are enormously elongated in caudal direction. Especially the Pars nasalis and Pars oralis pharyngis are expanded, the Tonsillae palatinae (see also Luckhaus 1969) are oblong and tube-shaped (Fig. 2.2) and the soft palate is elongated caudally. Therefore, the distance between the caudal end of the root of the tongue and the Aditus laryngis (Pars oralis pharyngis), is much, relatively, longer than in the *Felinae*. In the lion, the length of the Pars oralis pharyngis can reach up to more than 100 mm and the length of the soft palate more than 200 mm (e.g., 280 mm in the specimen dissected by Barone and Lombard 1967). In all examined species of the *Pantherinae* the free border of the elongated soft palate is located dorsal to the glottis. In the lion, the tiger and the jaguar an Arcus palatoglossus is lacking and the orifice between the oral and the pharyngeal cavity is solely and only slightly narrowed by the root of the tongue. Many elastic fibres can be observed within the pharyngeal walls. The widened pharyngeal cavity probably has evolved in adaptation to the intake and swallowing of large food items, whereas the enormous length of the pharynx in the examined species of *Pantherinae* is assumed to relate to their vocalizations. Most probably, the elongated pharynx 'tube' acoustically alters (filters) the source signal which is produced in the far caudally positioned larynx.

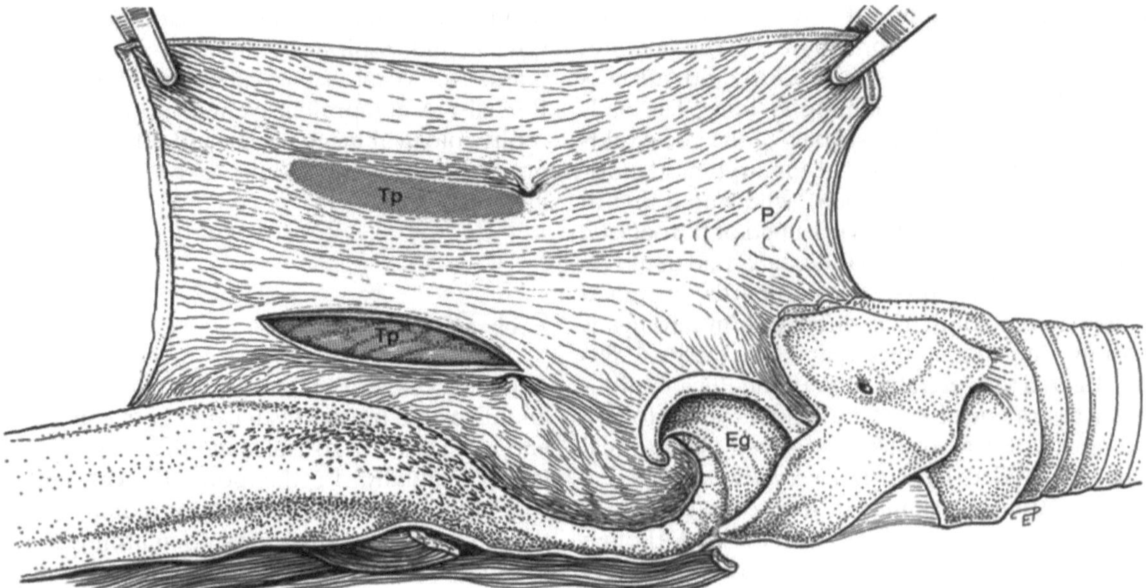

FIG. 2.2. Pars oralis pharyngis of the tiger. Left pharyngeal wall and soft palate, Velum palatinum (P) are cut. *Tp*, Tonsillae pharyngeae; *P*, soft palate; Velum palatinum (caudal part), *Eg*, Epiglottis.

2.4 Hyoid Apparatus

The hyoid apparatus supporting the larynx, pharynx and tongue consists of several elements. The basal element is the Basihyoideum (Corpus) with which the Ceratohyoideum (Cornu minus) and the Thyrohyoideum (Cornu majus) are connected. In all felid species examined, the basihyoid lies in a transverse plane. It is topographically adjacent to the larynx and therefore in some adult *Pantherinae* it is situated in the middle or caudal part of the neck. In species of the *Felinae* and in juvenile *Pantherinae* the basihyoid lies on a level with the skull base or the atlas. The suspensory apparatus of the hyoid which links the Ceratohyoideum with the skull base consists of the Epihyoideum, the Stylohyoideum and the Tympanohyoideum. In mammals, the suspensory apparatus of the hyoid varies with respect to the osseous, cartilaginous or ligamentous formation of its constituent parts. Members of the *Felinae* and the clouded leopard possess a suspensory apparatus the parts of which are ossified and/or cartilaginous (see also Owen 1835; Pocock 1916; Weissengruber et al. 2002). In the Pantherinae, with the exception of the clouded leopard, however, the epihyoid consists of an elastic ligament. In the lion, a small bone is embedded within the ligament. Pocock (1917) stated that this elastic ligament within the hyoid apparatus of almost all *Pantherinae* 'confers great mobility upon the larynx, which is not held close up to the base of the skull'. In our formalin-fixed specimens the length of the epihyoid ligament varies from 45 mm (tiger) up to 110 mm (lion), but in a fresh specimen of a lion it could be stretched to the length of more than 150 mm. The tympanohyoid consists of cartilage or calcified cartilage, it is situated next to the Bulla tympanica and syndesmotically attached to the skull. In the adult lion, jaguar and tiger, the ventral end of the tympanohyoid is rotated caudally about a dorsoventral axis. Since in juvenile *Pantherinae* this rotation is not discernible and the basihyoid together with the larynx is situated in a far rostral position comparable to that in *Felinae* (Fig. 2.3), it is assumed that a 'descensus' of the larynx and an accompanying rotation of the tympanohyoid occur during postnatal development of the mentioned species. In the Pantherinae, the stylohyoid is a slender, rod-shaped, slightly bent bone and its ventral end points caudally. In *Felinae* and the clouded leopard, the stylohyoid is oriented rostroventrally (Fig. 2.4). Contrary to other mammals such as artio-

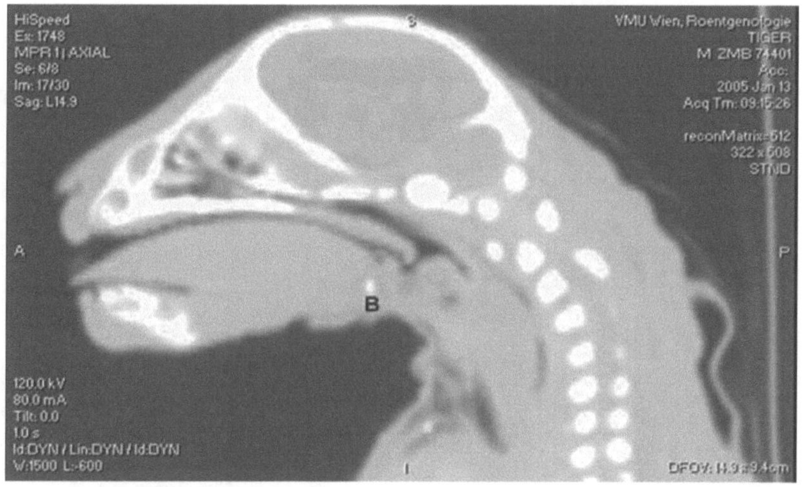

FIG. 2.3. CT image of the head and neck of a newborn tiger. *B*, basihyoid.

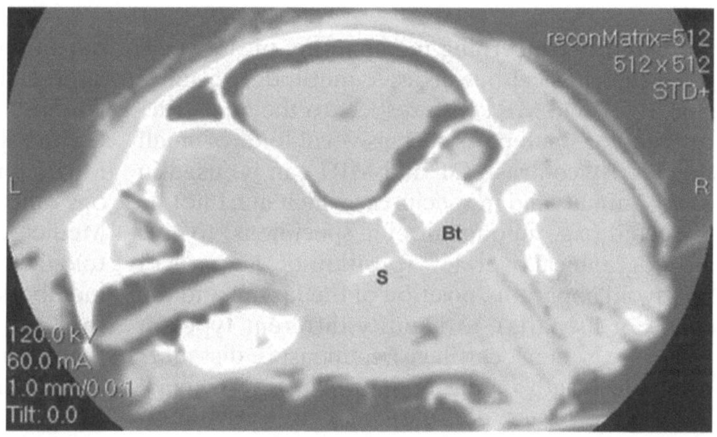

FIG. 2.4. CT image of the head and neck of an oncilla (*Leopardus tigrinus*). *S*, stylohyoid; *Bt*, Bulla tympanica.

dactyls or the horse (Schaller 1992) in which a massive stylohyoid occurs and the hyoid apparatus supports other anatomical structures (e.g., the tongue), the slender stylohyoid of *Pantherinae* simply serves as an 'extension'-element of the hyoid suspensory apparatus.

2.5 Position of the Larynx

Owing to the long epihyoid ligament, the rotated tympanohyoid, the caudoventrally oriented stylohyoid and the elongated pharynx, the larynx of adult *Pantherinae* (except that of the clouded leopard) is located on a level with the caudal cervical vertebrae or the first thoracic vertebra (Fig. 2.5). In one lioness examined, the larynx was situated within the thoracic inlet (Apertura thoracis cranialis). The clouded leopard has not been investigated yet, but it seems unlikely that it possesses a 'descended' larynx as the hyoid apparatus of this species is 'completely ossified'.

With the probable exception of the clouded leopard, the low-neck position of the larynx in *Pantherinae* seems to be permanent in the live animal. Interestingly, in the lion, the only species so far investigated in this respect, the long sternohyoid and sternothyroid muscles do not originate from the sternal manubrium as in other mammals (Schaller 1992). Instead, they originate from the inner surface of the chest wall close to the 3rd or 4th rib cartilage. This remarkable adaptation might involve the potential for further temporary retraction of the already far caudally positioned larynx of *Pantherinae* during vocalizations, reminding of the situation in red and fallow deer (Fitch and Reby 2001).

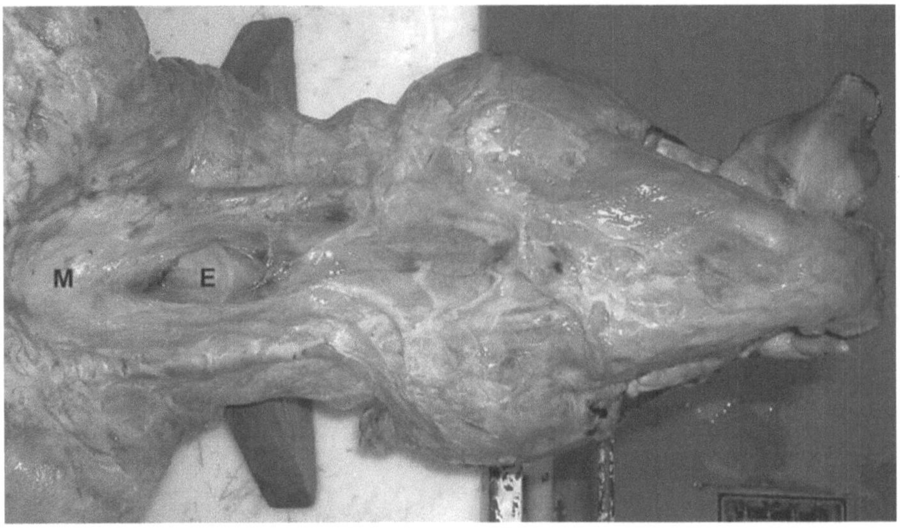

FIG. 2.5. Lion. Skin of head and neck removed, ventral view. *M*, Manubrium sterni; *E*, Epiglottis.

In studies concerning the anatomy of the vocal tract, the classical but invasive methods are superior to CT and MRI with respect to the fine structure and/or small parts such as vocal folds of small species. By contrast, CT and MRI can be used in living animals and in specimens which must not be dissected (e.g., unique museum specimens) to give a topographical overview (position of the parts of the hyoid apparatus, position of the larynx and the shape of its parts), to identify different types of tissues (e.g., bone, cartilage), to measure distances (e.g., the length of the upper vocal tract), signal intensities, tissue densities and to provide 3-dimensional reconstructions of distinct structures or cavities (e.g., shape of the pharyngeal or laryngeal cavities). In dead animals or specimens, the target does not move with every breath or during swallowing which might adversely affect image quality in clinical patients. Micro-CT may be used examining small structures or specimens. It is obvious that data from modern imaging techniques help enormously to link morphological findings with bioacoustical characteristics of vocalizations (e.g., vocal tract length with formant frequencies).

Acknowledgments

We thank Dr. Gustav Peters at the Zoological Research Institute and Museum Alexander Koenig, Bonn, Germany for long-time cooperation, access to many specimens, hospitality and outstanding support in felid bioacoustics. We appreciate cooperation with the Museum of Natural History, Humboldt-University Berlin, Germany. We owe thanks to Dr. Kirsti Witter and Magdalena Helmreich at the Institute of Histology and Embryology, University of Veterinary Medicine Vienna, Austria for keen support in histological processing. Technical support by the staff members of the Clinic for Diagnostic Imaging and the Institute of Anatomy is highly appreciated. We thank Dr. Eva Polsterer for graphical support and outstanding artwork.

References

Barone R, Lombard M (1967) L'appareil digestif du lion. Bull Soc Sci Vet Et Med Comparee Lyon 69:297–306

Fitch WT, Reby D (2001) The descended larynx is not uniquely human. Proc Royal Soc Lond B 268:1669–1675

Frazer Sissom DE, Rice DA, Peters G (1991) How cats purr. J Zool 223:67–78

Hast MH (1986) The larynx of roaring and non-roaring cats. J Anat 149:221–222

Hast MH (1989) The larynx of roaring and non-roaring cats. J Anat 163:117–121

Luckhaus G (1969) Vergleichend-anatomische Betrachtungen über den Pharynx des Löwen. Zentralblatt für Veterinärmedizin A 16:240–256

McElligott AG, Birrer M, Vannoni E (2006) Retraction of the mobile descended larynx during groaning enables fallow bucks (Dama dama) to lower their formant frequencies. J Zool 270:340–345

Owen R (1835) On the anatomy of the cheetah, Felis jubata, Schreb. Trans Zool Soc London 1:129–136

Peters G (1981) Das Schnurren der Katzen (Felidae). Säugetierkundliche Mitteilungen 29:30–37
Peters G (2002) Purring and similar vocalizations in mammals. Mammal Rev 32:245–271
Peters G, Hast MH (1994) Hyoid structure, laryngeal anatomy, and vocalizations in felids (Mammalia: Carnivora: Felidae). Z Säugetierkunde 59:87–104
Peters G, Tonkin-Leyhausen BA (1999) Evolution of acoustic communication signals of mammals: Friendly close-range vocalizations in felids. J Mammal Evol 6: 129–159
Pocock RI (1916) On the hyoidean apparatus of the lion (F. leo) and related species of Felidae. Ann Mag Nat Hist Zool Bot Geol 8:222–229
Pocock RI (1917) The classification of existing Felidae. Ann Mag Nat Hist 119:329–350
Schaller O (1992) Illustrated veterinary anatomical nomenclature. Ferdinand Enke, Stuttgart
Sunquist M, Sunquist F (2002) Wild cats of the world. The University of Chicago Press, Chicago, IL
Shiba K, Umezaki T, Zheng Y, Miller AD (1997a) The nucleus retroambigualis controls laryngeal muscle activity during vocalization in the cat. Exp Brain Res 115:513–519
Shiba K, Yoshida K, Nakajima Y, Konno A (1997b) Influences of laryngeal afferent inputs on intralaryngeal muscle activity during vocalization in the cat. Neurosci Res 27:85–92
Weissengruber GE, Forstenpointner G, Peters G, Kübber-Heiss A, Fitch WT (2002) Hyoid apparatus and pharynx in the lion (Panthera leo), jaguar (Panthera onca), tiger (Panthera tigris), cheetah (Acinonyx jubatus) and domestic cat (Felis silvestris f. catus). J Anat 201:195–209
Wilson DE, Reeder DM (2005) Mammal species of the world. Johns Hopkins University Press, Baltimore, CT

3 The Anatomical Foundation for Multidisciplinary Studies of Animal Limb Function: Examples from Dinosaur and Elephant Limb Imaging Studies

John R. Hutchinson[1], Charlotte Miller[1], Guido Fritsch[2], and Thomas Hildebrandt[2]

3.1 Introduction

What makes so many animals, living and extinct, so popular and distinct is anatomy; it is what leaps out at a viewer first whether they observe a museum's mounted *Tyrannosaurus* skeleton or an elephant placidly browsing on the savannah. Anatomy alone can make an animal fascinating – so many animals are so physically unlike human observers, yet what do these anatomical differences mean for the lives of animals?

The behavior of animals can be equally or more stunning- how fast could a *Tyrannosaurus* move (Coombs 1978; Alexander 1989; Paul 1998; Farlow et al. 2000; Hutchinson and Garcia 2002; Hutchinson 2004a,b), or how does an elephant manage to momentarily support itself on one leg while 'running' quickly (Gambaryan 1974; Alexander et al. 1979; Hutchinson et al. 2003, 2006)? Yet to attain a fundamental understanding of many animal behaviors we must venture beneath their skin surface and analyze the anatomical structures that constrain and enable visible behaviors. An elephant does not manage the aforementioned unipodal feat by sheer willpower or neurological activity alone – muscles, tendons, and bones must provide enough support. Indeed, in the case of an extinct animal, except for fossilized footprints and other tantalizing evidence of behavior, anatomy is all we have to work with, at first. Yet that does not mean that behavior cannot be addressed by indirect scientific means.

Here we use two intertwined case studies from our research on animal limb biomechanics, one on extinct dinosaurs and another on extant elephants, to illustrate how anatomical methods and evidence are used to solve basic questions. The dinosaur study is used to show how biomechanical computer modeling can reveal how extinct animal limbs functioned (with a substantial margin of error that can be addressed explicitly in the models). The elephant study is used to show how classical anatomical observation and three-dimensional (3D) imaging have powerful synergy for characterising extant animal morphology, without biomechanical modeling, but also as a first step toward such modeling.

We focus on how a combination of classical techniques (particularly dissection, osteology, and functional anatomy) and modern techniques (especially imaging and computer modeling) are integrated to reveal how limbs function, and how anatomy is related to behavioral analyses such as biomechanics. We hope that this might inspire other functional biologists, clinicians, or researchers in other disciplines to see how an integration of anatomical methods can yield exciting insights into animal function or behavior.

[1]Structure and Motion Laboratory, Department of Veterinary Basic Sciences, The Royal Veterinary College, University of London, Hatfield, Hertfordshire AL9 7TA, UK, and [2]Leibniz Institute for Zoo and Wildlife Research (IZW), Berlin, Germany

3.2 Example 1: The Dinosaur *Velociraptor mongoliensis* Osborn 1924

Velociraptor (e.g., of *Jurassic Park* media fame) is a member of the theropod dinosaurs: habitually bipedal, ancestrally carnivorous animals that include living birds (e.g., Gauthier 1986; Sereno 1999). It is a Late Cretaceous representative of the closest known relatives of all birds: the deinonychosaurs. Although it is a late branch of this lineage and slightly large (<25 kg; e.g., Hutchinson 2004b) compared with earlier members (e.g., *Microraptor*; Xu et al. 2003) it is useful for understanding the evolution of traits present in extant birds, such as crouched striding bipedalism or the functions of homologous muscles in standing and moving (e.g., Gatesy 1990; Carrano 1998; Hutchinson and Gatesy 2000). As a model of the contemporary *Tyrannosaurus rex*, from a phylogenetically more distant lineage of theropods, is available (Hutchinson et al. 2005), we had a guide for how to construct this model and also a second extinct taxon to which we could compare the mechanics of homologous muscles.

We used a Cyberware 3030 3D color laser digitizer (Cyberware, Inc., Monterey, California, USA) to capture the 3D surface geometry of the pelvis and hindlimb of two similarly-sized specimens (Institute of Geology, Mongolia specimens IGM 100/986 and 100/985). Neither was 100% complete and intact, so we had to scale elements from the better-preserved pelvis of the smaller specimen (IGM 100/985) to match the larger, by using linear measurements (lengths and heights) of the two to establish proportional transformations. After we filled any holes and deformations were corrected for the best-preserved right or left elements, we 'mirrored' these elements about their midline axis in order to create the element of the opposite side of the limb. Although we created a left limb (and left half of the pelvis) for simplicity here we focus only on the single right limb and right side of the pelvis. Unlike elephants, *Velociraptor* was habitually bipedal, so we do not focus on its forelimb anatomy and mechanics here. We took additional measurements of the tail and other adjacent body parts to aid placement of soft tissues even though these elements were not included in the model.

We articulated these 3D images of our composite *Velociraptor* into a poseable skeleton in SIMM musculoskeletal modeling software (Musculographics, Inc.; Chicago, IL) following the methods of Hutchinson et al. (2005). Briefly, we used hemispheres centered on the acetabulum and femoral head to calculate the geometric origins (0,0,0 in x,y,z coordinates) of the pelvis and femur segments to articulate them. We calculated the approximate geometric origins of the femoral condyles, tibial malleoli, and distal metatarsal condyles in order to establish the mean centers of joint rotation of their respective segments (tibiotarsus, metatarsus, and pes) as well. The lengths of the femur, tibia, and metatarsus were increased by 7.5%, 5%, and 10% to account for missing soft tissue between bony joint surfaces (Hutchinson et al. 2005). These procedures allowed articulation of the skeleton in the software (Fig. 3.1). Manual manipulation of joint orientations allowed us to check whether the movements of the segments were realistic throughout the ranges of motion allowed in the model, which were assessed by examination of the original material to bound ranges of feasible joint flexion/extension, ab/adduction, and lateral/medial rotation.

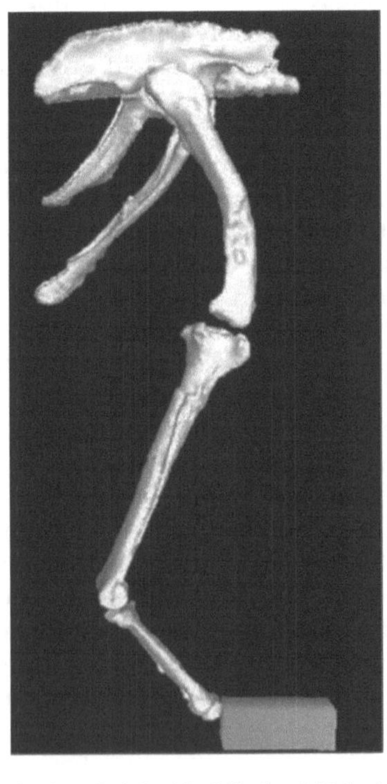

Fig. 3.1. Articulated right hind limb of *Velociraptor mongoliensis* in musculoskeletal modeling software, in lateral view. The pose shown is an arbitrary one; the reference pose was with all bones above the foot fully vertical.

Subsequently, we used muscle scars and comparison with dissections of extant relatives of *Velociraptor* (birds, crocodilians and other Reptilia) to reconstruct the hindlimb musculature of the right leg, following Hutchinson (2001a,b, 2002; Hutchinson and Gatesy 2000; Carrano and Hutchinson 2002). We then added curved wrapping surfaces to guide muscles on more realistic non-linear 3D paths, again using the skeletal anatomy and extant taxa as a guide (Hutchinson et al. 2005). Once we checked these muscle attachments to ensure no anomalous results occurred (e.g., sudden jumps or loops of the muscle-tendon unit paths as joints were moved), the model was ready (Fig. 3.2).

Here we focus on the relationships of muscle moment arms (see An et al. 1984; Delp et al. 1999) to joint angles (as per Hutchinson et al. 2005) for the major flexors and extensors of the hip joint, as these muscles are presumed to have undergone the most dramatic anatomical and functional transformations between the common ancestors of tyrannosaurs (basal coelurosaurian theropods; distant from birds) and deinonychosaurs (eumaniraptoran theropods; sister group [i.e., cousin lineage] to birds). We compared our model of *Velociraptor* with a model of *Tyrannosaurus* (Hutchinson et al. 2005), with muscle moment arms normalised by a reference bone length to

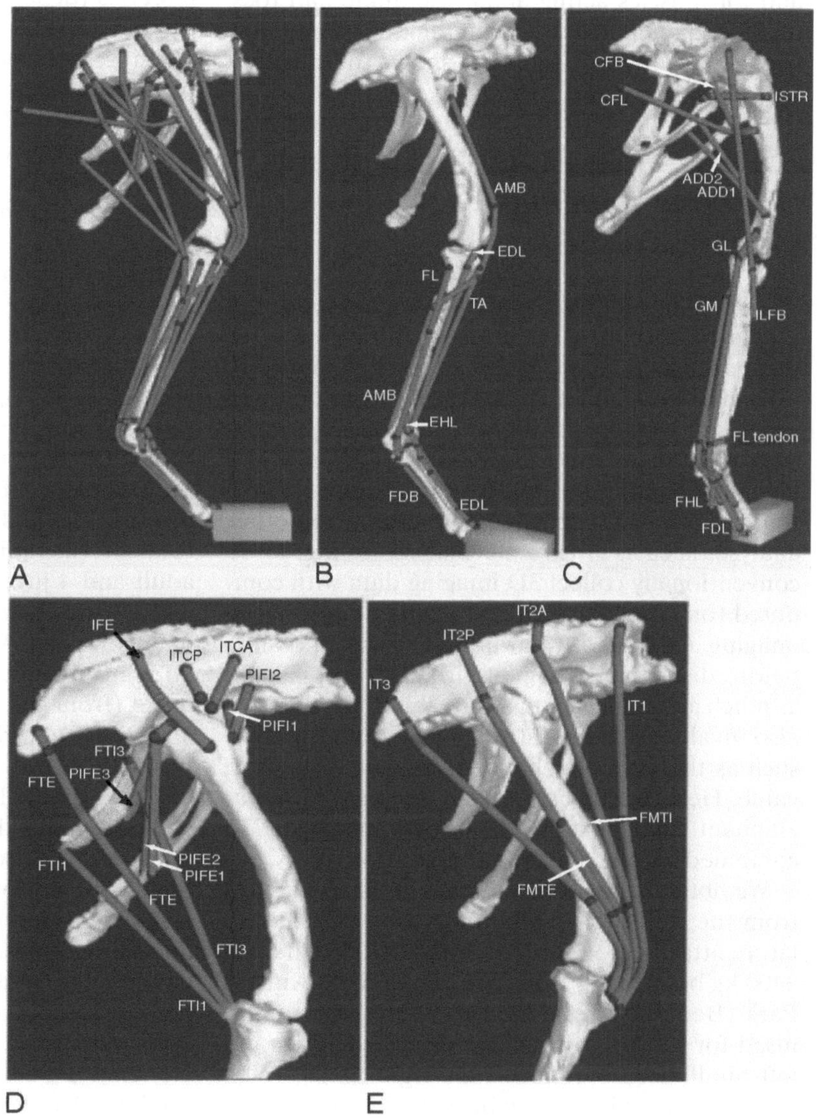

FIG. 3.2. Muscles (*red lines*) of *Velociraptor* right hind limb model, in lateral view with all shown (**A**), oblique craniolateral view of lower limb muscles (**B**), oblique caudolateral view of some hip extensor and lower limb muscles (**C**), dorsolateral view closeup of hamstring (flexor cruris complex), external puboischiofemoral, and deep dorsal thigh muscles (**D**), and craniolateral view closeup of M. iliotibialis and M. femorotibialis knee extensor muscles (**E**). See Table 3.2 for muscle abbreviations (see Color Plates, Fig. 3.2).

TABLE 3.1. Reference lengths of segments (from musculoskeletal model) used for normalizing moment arm measurements. The 'Ratio' column shows the ratio of the length in *Tyrannosaurus* to the corresponding length in *Velociraptor*.

Length	*Velociraptor*	*Tyrannosaurus*	Ratio
Femur	0.160	1.22	7.63
Tibiotarsus	0.221	1.27	5.75
Tarsometatarsus	0.120	0.780	6.50

remove the confounding effects of body size. For muscles acting about the hip joint, the femur length (hip to knee joint distance) was used as the reference length; for muscles acting about the knee, the 'tibiotarsus' length (knee to ankle joint), and for muscles acting about the ankle and toes, the 'tarsometatarsus' length (ankle to metatarsophalangeal joint). Table 3.1 shows these values for *Velociraptor* and *Tyrannosaurus*.

3.3 Example 2: Asian Elephants (*Elephas maximus* Linnaeus 1758)

As elephant anatomy is poorly understood and access to specimens is severely difficult due to their protected conservation status, low mortality rates and population sizes, remote locations, and size (e.g., handling, transport, and storage logistics), making the most of available specimens is a high priority for our research. Dissection of course renders cadaveric specimens unusable for other analyses such as *in vitro* mechanical testing, so we conventionally collect 3D imaging data with computed tomography (CT) and magnetic resonance imaging (MRI; not shown here) when possible before dissecting specimens, in order to preserve as much anatomical data as possible. Such data are also invaluable for constructing computer models such as the example above, so imaging has much value. Here, we focus on what we learned about elephant limb anatomy from 3D imaging (CT) combined with conventional dissection.

We obtained the fore and hind limbs (separated from the torso/pelvis, with most extrinsic musculature attached) of an adult female (~25 years old; 3400 kg body mass) from Whipsnade Wild Animal Park (Bedfordshire, UK). This animal was euthanised for severe arthritis and osteomyelitis in the left hindlimb; we obtained the right limbs which appeared to be in reasonably sound musculoskeletal condition, although we make some notes on evident pathologies below.

We did CT scanning on a Picker PQ5000 station (General Electric Company, Coventry, UK), with typically 150 axial x-ray slices at 1.5–5 mm intervals (depending on specimen size), 512 × 512 pixels, 120 kvP and 200 mA intensity. After imaging, we loaded the raw DICOM format scanner images into MIMICS 10.1 (Materialise, Inc; Leeuwen, Belgium) software where we used gray-value thresholding to identify tissues of different densities (e.g., muscle, tendon, and bone) and create 3D images of each identified structure (Fig. 3.2). These 3D models are the main subject of this part of the study.

After imaging, we quickly dissected all segments and stored them in a freezer (–20°C) or cold room (4°C) when not in use. Abundant digital photos and video were taken as supplementary documentation during dissection. We used supplementary cadaver data (collected in the same fashion) from the complete fore and hindlimbs of one neonatal, one juvenile and one adult Asian elephants plus a total of (including from the complete limbs) 5 adult and 4 juvenile/neonatal manus, and 4 adult and 4 juvenile/neonatal pedes (of which one manus was an African adult, and one pes was an African juvenile), plus dry museum bone specimens (from the Cambridge Museum of Zoology and Royal Veterinary College Anatomy Museum) to compare with our main specimen in order to assess individual variation or pathology.

Additionally during the dissections we collected data on the muscle-tendon unit architecture of all limb muscle groups, following Payne et al. (2004, 2005) and references therein. This included muscle and tendon masses and lengths, and muscle fascicle lengths and pennation angles, with which muscle force production can be estimated (e.g., Alexander et al. 1981; Zajac 1989; Alexander and Ker 1990; Payne et al. 2004, 2005). We do not focus on these data here, but they are essential anatomical data needed for biomechanical analysis and few such data exist for most extant taxa. A comprehensive, quantitative anatomical study would include all such measurements in addition to traditional origins, paths, and insertions.

Here, we focus mainly on the osteological anatomy of Asian elephants (which unlike *Velociraptor* or African elephants, is not well described in the literature), using our adult female specimen as a primary example but with ancillary data from other specimens as noted above, and soft tissue data where most relevant for interpreting the osteology.

3.4 Results

3.4.1 Velociraptor *Model*

Absolute values for hip muscle moment arms in the sagittal plane are in Table 3.2. Generally the model's flexor and extensor muscle moment arm patterns (Figs 3.3,3.4) matched those of *Tyrannosaurus* (Hutchinson et al. 2005). For the hip flexor and extensor muscles, most muscles had relatively larger moment arm values in *Tyrannosaurus*. Some relative magnitudes were still strikingly different, however: the deep dorsal thigh muscles (M. iliotrochantericus caudalis and M. puboischifemoralis internus; both parts of each group) showed reduced hip flexor moment arms and switches between hip flexor and extensor moment arms in relatively extended femoral orientations in *Velociraptor*, unlike in *Tyrannosaurus* (Fig. 3.4). In contrast, M. flexor tibialis internus 1 (dubiously present; Carrano and Hutchinson, 2002) had relatively larger moment arms in *Velociraptor* (Fig. 3.3E). M. puboischiofemoralis externus 1 and 2, whose origins were rotated caudally in maniraptoran theropods with retroversion of the pubic bones, showed relatively much lower moment arms for hip flexion, but still were competent hip flexors. Lower limb muscles were relatively similar between the two taxa, which is consistent with the observation that lower limb anatomy did not transform as drastically between basal coelurosaurs and eumaniraptoran theropods (e.g., Hutchinson 2002).

3.4.2 Elephant Models

Our rendered 3D models of elephant limb bones were of a high enough resolution that all major

TABLE 3.2. Muscle moment arm results for *Velociraptor* model. Muscle names, abbreviations, and moment arms at 0° and 45° hip joint angles (larger values are more flexed; 0° is a femur held perpendicular to the pelvic axis), and minimum, maximum, and mean values are shown. Moment arm units are in centimeters; negative moment arms are hip flexor whereas positive are extensor.

Muscle	Abbrev.	At 0°	At 45°	Min	Max	Mean
M. iliotibialis 1	IT1	−6.6	−4.8	−8.4	−4.8	−6.8
M. iliotibialis 2, preacetabular	IT2A	−4.0	−4.2	−4.2	−3.8	−4.1
M. iliotibialis 2, postacetabular	IT2P	−0.38	−0.65	−0.65	−0.29	−0.37
M. iliotibialis 3	IT3	5.4	6.0	3.3	6.1	5.0
M. ambiens	AMB	−3.1	−4.1	−4.1	2.6	−1.8
M. iliofibularis	ILFB	3.6	4.3	3.2	4.3	3.6
M. iliofemoralis externus	IFE	0.40	1.1	−1.9	1.1	−0.25
M. iliotrochantericus caudalis, anterior	ITCA	−1.3	−1.6	−1.6	0.87	−0.80
M. iliotrochantericus caudalis, posterior	ITCP	−1.4	−1.6	−1.5	0.59	−0.93
M. puboischiofemoralis internus 1	PIFI1	−2.3	−1.6	−2.6	−1.6	−2.3
M. puboischiofemoralis internus 2	PIFI2	−2.4	−2.0	−2.4	1.7	−1.5
M. flexor tibialis internus 1	FTI1	10	−2.4	−2.4	11	8.0
M. flexor tibialis externus	FTI3	3.0	0.39	0.39	3.5	2.7
M. flexor tibialis internus 3	FTE	6.4	4.3	4.3	7.3	6.5
M. adductor femoris 1	ADD1	3.9	−1.1	−1.1	5.0	3.4
M. adductor femoris 2	ADD2	3.9	0.25	0.25	4.4	3.4
M. puboischiofemoralis externus 1	PIFE1	−0.77	−1.7	−1.7	0.13	−0.70
M. puboischiofemoralis externus 2	PIFE2	−0.70	−1.4	−1.4	0.12	−0.58
M. puboischiofemoralis externus 3	PIFE3	−0.39	−0.66	−0.66	0.37	−0.24
M. ischiotrochantericus	ISTR	0.43	−0.41	−0.41	0.76	0.39
M. caudofemoralis brevis	CFB	4.5	4.5	3.0	4.7	4.1
M. caudofemoralis longus	CFL	4.4	3.2	3.2	5.9	4.7

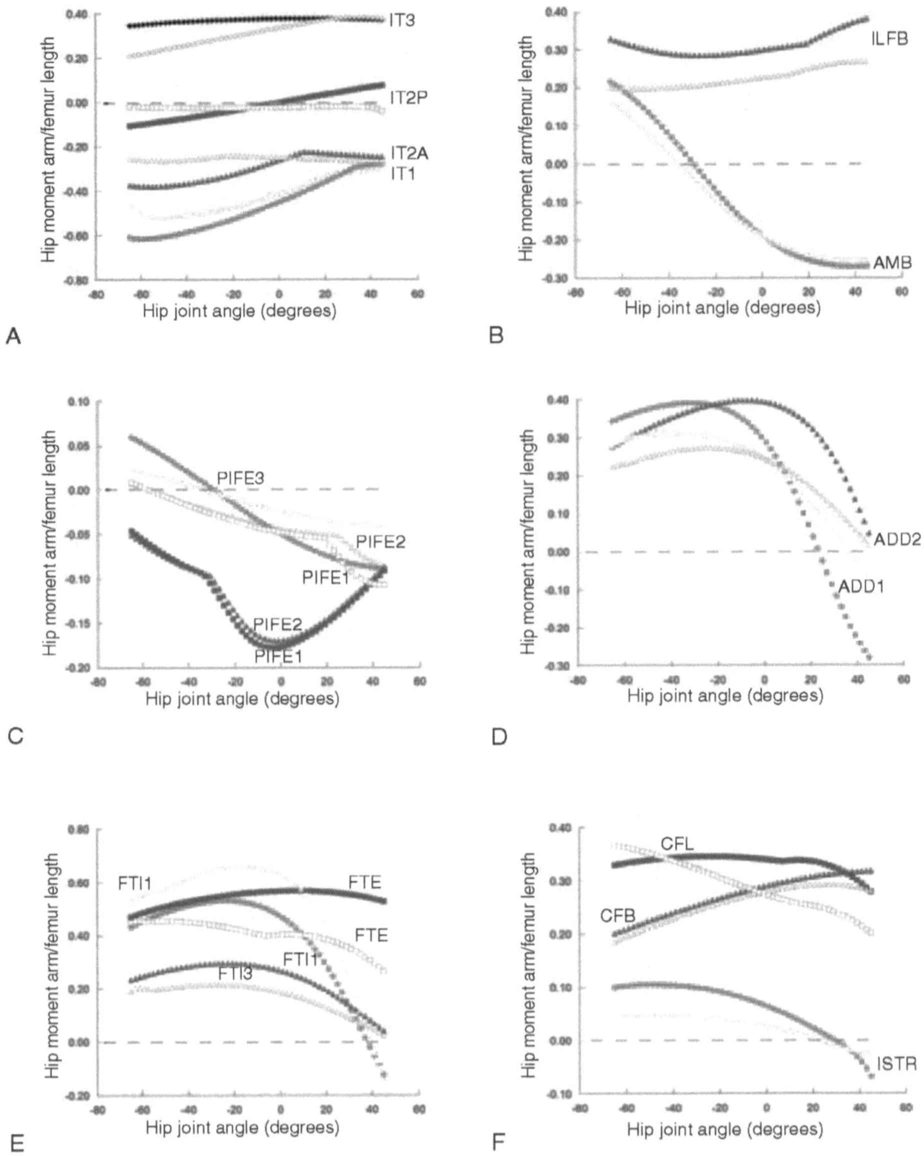

FIG. 3.3. Hip flexor/extensor moment arm-hip joint angle relationships for major hip muscles in *Velociraptor* (*open symbols*) and *Tyrannosaurus* (*filled symbols*). Moment arms are normalized by femur length (see text and Table 3.1) for size-independent comparisons. Positive moment arms are extensor; negative are flexor. A hip joint angle of 0° represents a vertical femur; negative values are extension past that point and positive values are flexion (femoral protraction). **A**, M. iliotibialis 1 – 3 (IT1 – 3). **B**, M. iliofibularis (ILFB) and M. ambiens (AMB). **C**, Mm. puboischiofemorales externi 1 – 3 (PIFE1 – 3). **D**, Mm. adductores femorii 1 + 2 (ADD1 + 2). **E**, Mm. flexores tibiales interni 1 + 3 (FTI1 + 3) et externus (FTE), and **F**, M. ischiotrochantericus (ISTR) and Mm. caudofemorales brevis et longus (CFB, CFL) (see Color Plates, Fig. 3.3).

anatomical features of the bones – even subtle muscle scars – described in the excellent osteological studies by Smuts and Bezuidenhout (1993, 1994) could be identified in our main specimen and other elephants we studied. We found few striking osteological differences between the literature descriptions of African elephants and our Asian elephants. Our specimens showed the same sexual dimorphism observed by the latter authors (e.g., more prominent muscle scars in female elephants; surprising considering the larger mean size of males). We proceed from proximal to distal elements of the fore and hind limbs (Fig. 3.5), noting where our specimen (or any others)

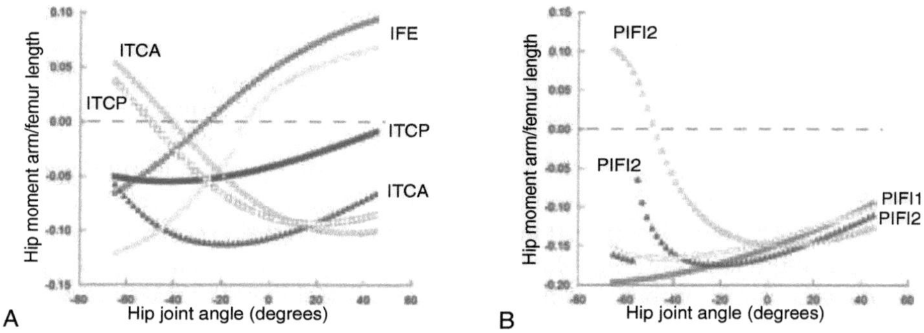

FIG. 3.4. Additional hip moment arm-joint angle relationships as in Fig. 3.3. **A**, M. iliofemoralis externus (IFE) and M. iliotrochantericus caudalis (ITCA, ITCP). **B**, Mm. puboischiofemoralis internus 1 + 2 (PIFI1 + 2) (see Color Plates, Fig. 3.4).

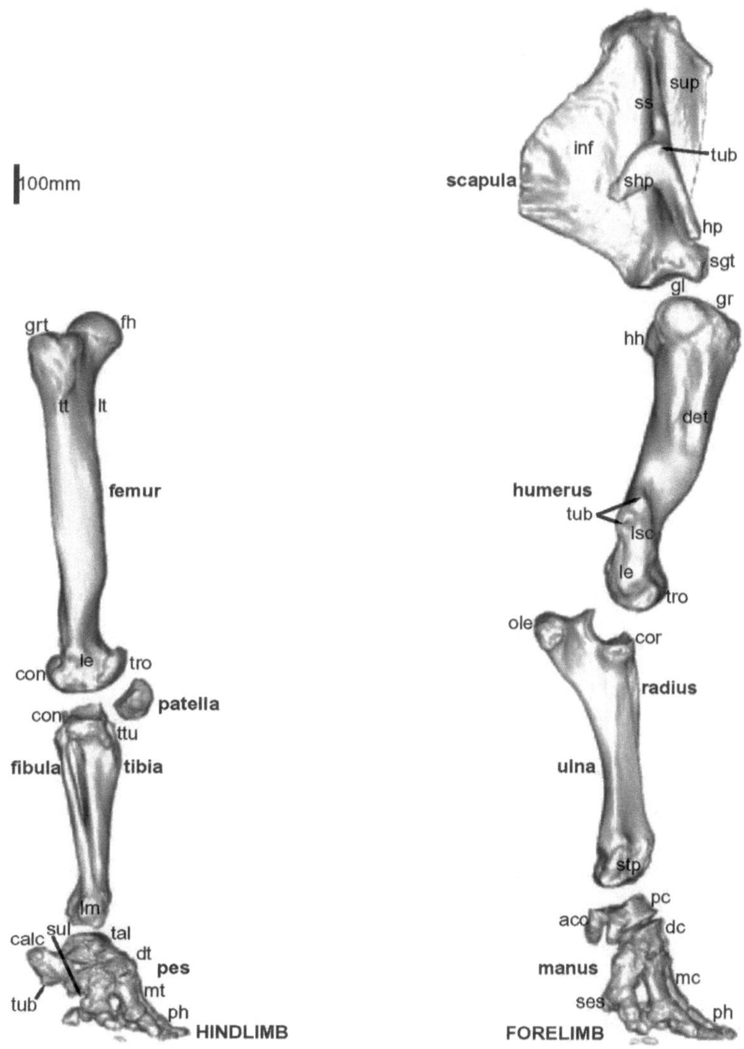

FIG. 3.5. Bones of the right fore and hind limb of our study specimen of *Elephas maximus*, three-dimensionally rendered in MIMICS software from CT scans. In lateral view. The major bones are moved slightly out of articulation so more of their surfaces can be seen. Abbreviations: acc = accessory carpal, calc = calcaneus, con = condyle, cor = medial coronoid process, dc = distal carpals (I-V), det = deltoid tuberosity, dt = distal tarsals (I-V), fh = femoral head, gl = glenoid surface, gr = greater tubercle, grt = greater trochanter, hh = humeral head, hp = hamate process, inf = infraspinous fossa, le = lateral epicondyle, lm = lateral malleolus, lsc = lateral supracondylar crest, lt = lesser trochanter, mc = metacarpals (I-V), mt = metatarsals (I-V), ole = olecranon process, pc = proximal carpals (I-V), ph = phalanges (I-V), ses = sesamoid(s), sgt = supraglenoid tubercle, shp = suprahamate process, ss = scapular spine, stp = styloid process, sul = sulcus, sup = supraspinous fossa, tal = talus, tro = trochlear surface, tt = third trochanter, tub = tubercle, ttu = tibial tuberosity (see Color Plates, Fig. 3.5).

showed important differences from Smuts and Bezuidenhout's (1993, 1994) African elephant specimens. All bony epiphyses were fully fused except where noted; our main specimen was almost skeletally mature (Roth 1984).

The scapula had three tubercles on the scapular spine region: one midway along the lateral surface of the spine (part of M. trapezius insertion), a second on the spine at the juncture of the hamate and suprahamate processes (probable origin of fascial sheet extending down the forelimb), and a third on the lateral tip of the hamate process (end of deltoid partial origin, and attachment for cranio-lateral and cranial 'lacertus fibrosus' fascial sheets).

The humerus revealed two distinct features of the crista supracondylaris lateralis not noted elsewhere: there was one quite large tubercle (with a distal invagination) on the caudolateral side of the crest near its proximal end (associated with the origin of M. anconeus), and then a second, smaller tubercle proximal to it on the crest (part of the M. brachioradialis origin).

The radius and ulna of our main specimen were fused together proximally quite strongly as well as partly fused at their distal articulation. The ulna had marked muscle scarring on the proximal half of the caudomedial surface of the corpus ulnae (biceps brachii partial insertion, as well as a second patch of scarring also visible in CT images on the proximal olecranon's caudolateral surface, distal to the tuber olecrani proper (part of the M. anconeus insertion).

The manus (Figs 3.5,3.6) showed strong ossification and rugosities but no evident pathology. We found no clear articulation of the ulnar carpal with metacarpal V – indeed, Smuts and Bezuidenhout's (1993) figures concur with this, unlike the text. The

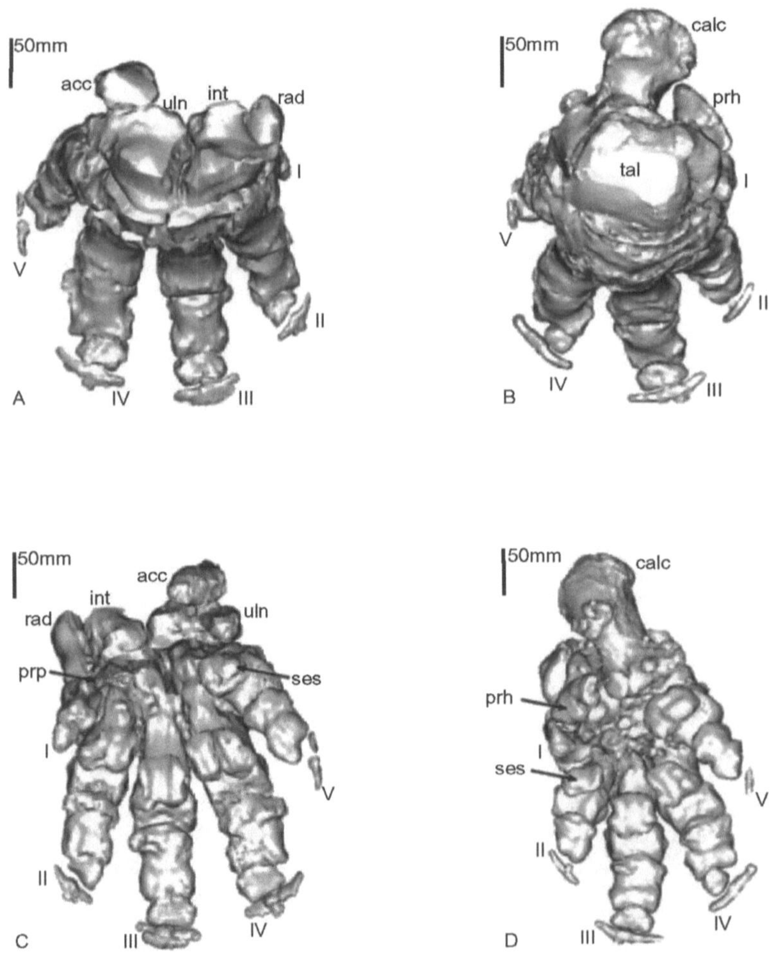

FIG. 3.6. Articulated bones of the right manus (**A**, **C**) and pes (**B**, **D**), in palmar (**A**, **B**) and planter (**C**, **D**) views. Abbreviations as in Fig. 3.5 caption plus: I-V = digits I-V, int = intermediate carpal, prh = prehallux, prp = prepollex, rad = radial carpal, tal = talus, uln = ulnar carpal (see Color Plates, Fig. 3.6).

accessory carpal was the normal insertion point for M. extensor carpi ulnaris (and M. extensor digitorum lateralis). Carpals 3 and 4 had plantar tuberosities, presumably for the insertion of M. flexor carpi radialis. The metacarpals match the descriptions of Smuts and Bezuidenhout (1993) very well, except that metacarpal II's proximal tuberosity was expanded proximally into a spine (this was not present in other individuals).

All manus digits except I had two large proximal sesamoids. Close examination of our main specimen showed at least one, maybe two ossifications in the location of the proximal sesamoid of digit I (Fig. 3.7A,B). However in our other five adult specimens we saw a single sesamoid in digit I in two specimens, two separate sesamoids in one specimen, and an ambiguous number (one or two but resolution too coarse to determine) in the others. Juveniles tended not to have sesamoids visible at all in digits I and V. We infer that at least Asian elephants have two sesamoids in digit I, but they may fuse or be indistinguishable in adults.

In our main specimen, digit I of the manus had only one phalanx (although in two adult inviduals we saw two phalanges), digits II-IV had the typical three phalanges (except one pathological adult individual only had two for each digit, and juveniles had anywhere from 2–3 identifiable phalanges, tending to increase with age as expected), and digit V had proximal and distal phalanges with a

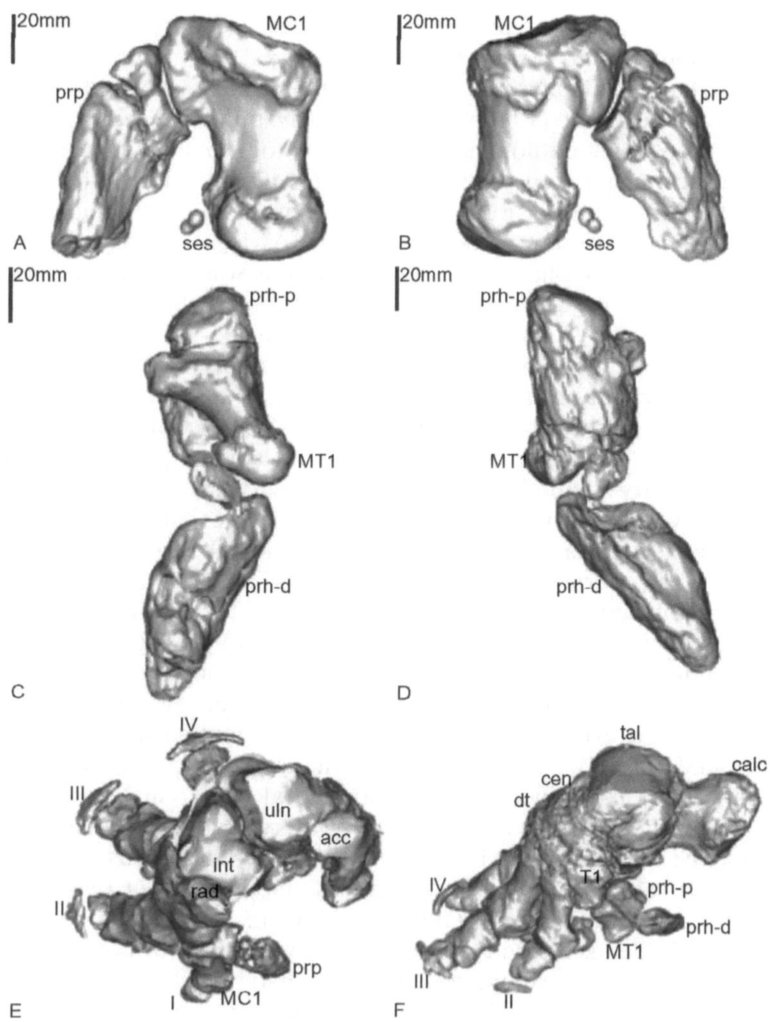

FIG. 3.7. 'Predigits' of the right manus and pes: prepollex in medial (**A**) and lateral (**B**) views, prehallux in cranial (**C**) and caudal (**D**) views, and dorsomedial views (not to scale) of manus (**E**) and pes (**F**), with articulations shown. Abbreviations as in Figs 3.5, 3.6 captions plus: cen = central tarsal, MC1 = metacarpal I, MT1 = metatarsal I, prh-d = distal segment of prehallux, prh-p = proximal segment of prehallux, T1 = (distal) tarsal I (see Color Plates, Fig. 3.7).

gap separating them (i.e., no ossified middle phalanx; contrary to the findings of Smuts and Bezuidenhout (1993), who found no ossified ungual). On closer investigation of the CT images, there appeared to be a cartilaginous middle phalanx of digit V that had failed to ossify in our main specimen, and this was likewise common but individually variable in other Asian elephants – two of our larger adult individuals even had a complete set of three ossified phalanges. Digits III and IV had ungual (distal) phalanges that, unlike the specimens described by Smuts and Bezuidenhout (1993), had articulations evident with the middle phalanges, whereas digit II had the usual gap between these two distalmost phalanges. These two patterns were variably present in other specimens. Digit V's ungual phalanx was divided into medial and lateral halves; likewise variable even among adults. Below we add additional notes on the anatomy of the prepollex.

We observed no other external features of the femur, patella, tibia, or fibula (Fig. 3.5) that Smuts and Bezuidenhout (1994) did not already point out, except some pathologies noted below for our main specimen. Likewise we saw the same details that the latter authors note; there were no marked osteological differences between their African and our Asian elephants. The tibial epiphysis was incompletely fused, unlike the other limb bones; although this is slightly unusual for typical elephants (Roth 1984) we assume that it is merely individual variation rather than pathology.

Some details of pes osteology (Figs 3.5,3.6) in our main specimen were obscured by some extraordinary pathologies (described further below), yet the basic anatomy of elephant pedes was still recognizable, especially distally. The calcaneus had marked ligamentous and tendinous scarring, including on the plantar surface of the calcaneal tuber, corresponding to the adductors and abductors of digit V. Additionally, there was a strong lateral tuberosity at the base of the tuber, caudal to the articulations with metatarsal V and tarsal IV. This tuber also bounds a tendinous sulcus with a hooklike project of the caudolateral base of the articular surface for metatarsal V; this sulcus is for the tendon of M. tibialis caudalis. Talus and distal tarsal anatomy was also as described by Smuts and Bezuidenhout (1994), with all major details visible except where obliterated by pathological fusion or osteophytes. The pes had the typical five metatarsal bones, conforming quite well to the description by Smuts and Bezuidenhout (1994). Pathologies, especially proximally, are noted below.

As in African elephants, the hallux (digit I) had only one very small, flat proximal sesamoid (lying between the metatarsal trochlea and the prehallux; described below) and no phalanges. This was variable in our eight other specimens – adults and immature elephants alike had either zero or one sesamoid. Curiously, all five of our adult specimens lacked hallucal phalanges whereas all four juveniles had one phalanx, so phalanx number is either individually or ontogenetically variable (our sample size is too small to determine). The other digits each had two proximal sesamoids (none fused as Mariappa 1986 claimed for digit II).

Digits II and V had two ossified phalanges, lacking a middle phalanx – the proximal and distal (ungual) phalanges were separated by a gap where a presumably cartilaginous nodule represented the middle phalanx. However this pattern differed in adults: three complete digits were present in digit II in the four other individuals, whereas half had either two or three phalanges in digit V. Juveniles had two ossified phalanges in digit II and 1–2 in digit V. The unguals of these two digits were typically simple and crescent-shaped; ungual II had a hooked ventromedial tuberosity, whereas ungual V lacked any marked tuberosity in our main specimen, and this shape was likewise variable among other specimens. Digits III and IV usuallly had the usual full complement of three phalanges in adults (but one had two phalanges in digit IV), whereas juveniles had 2–3. If an elephant had three phalanges in any one digit, it was always digit III. The ungual of digit III had a tuberosity that was offset slightly medially (unlike in typical African elephants), whereas the tuberosity on digit IV was markedly reduced, and the middle phalanx was much smaller than in digit III. However, this asymmetry was not typical for other specimens.

Our CT images of our main specimen and others showed some remarkable features of the fore and hind feet that deserve greater attention, so we focus on these here. From our observations of the feet of four African elephants (from juvenile to adult) we are confident that these comments apply to both major species of elephants. The anatomy of the bizarre 'predigits' (prepollex and prehallux; digit-like structures of unknown composition or ontogeny that lie medial to digit I) of elephant manus and pedes caused some confusion in early literature concerning the number of digits in elephants (see Smuts and Bezuidenhout 1993, 1994). This anatomy has not previously been described in detail, nor has its ontogenetic variation been inspected. As CT scanning provides noninvasive anatomical data on the *in situ* positions and structure as well as relative density (i.e., calcification) of these mysterious elements, here we add detailed description.

The prepollex (Fig. 3.7A,B,E) is a single segment (in all specimens) that articulates only with metacarpal I. It is claw or hook-shaped, with the concave surface of the hook oriented craniomedially. Its articulation is a large rugose, convex boss on the proximal caudal (plantar) surface of metacarpal I, forming a saddle-shaped joint articulation with the concave, crescent-shaped proximal surface of the prepollex. The prepollex has a longitudinal ridge on its caudolateral and craniomedial surfaces.

The prehallux (Fig. 3.7C,D,F) is a slightly curved, rodlike structure that articulates with tarsal I and metatarsal I, on their caudal (plantar) surfaces. The prehallux consists of at least two and sometimes three calcified segments (variable among specimens; Neuville (1935, fig. 26) showed four possible centers of calcification) that have an articulation at the level of the distal end of metatarsal I. A concave notch is present on tarsal I that a rounded peg on the proximal segment of the prehallux fits into, approximating a ball-and-socket joint. This prehallucal segment then lies parallel to metatarsal I and sits appressed to a flat caudomedial groove in that bone (Fig. 3.7F). The nature of the bones surrounding it, especially tarsal I's main articulation with it, is that the movements of the prehallux are restricted to a dorsocaudal plane.

Both 'predigits' calcify (or ossify; histological data are needed to test this) throughout growth, with the calcification of cartilages proceeding from proximal and distal toward the middle, and from the surfaces inward toward the core, with a tendency for thicker calcification on the lateral sides.

We also observed interesting pathologies in our primary female specimen that we are not aware of ever having been described in elephants, even though the disorders (osteomyelitis and osteoarthritis, primarily) they are symptoms of are major causes of mortality in captive elephants (Csuti et al. 2001). These pathologies were centered around the tarsus (Figs 3.5,3.6). The distal end of the tibia (caudal side of the cochlea tibiae) had extended further distally behind the talus, forming a prong that stiffened the back of the ankle joint. Likewise the distal fibula showed excessive growth, and the talus had enlarged proximal articular facets, especially laterally.The tarsals show other pathologies related to decreasing the mobility of the ankle joint complex. The distal tarsals (centrale and tarsals I-IV) were largely fused to each other, sharing marrow cavities. Likewise, the medial and distal surfaces of the calcaneus was fused to the talus, centrale, and tarsal IV. Generally, the surfaces of all tarsals was roughened by this ankylosis and scarred with pits, vascular channels, and osteophytes (spurs). This pathology extended into the proximal metatarsals, with signs of fusion particularly between metatarsal V and tarsal IV, and metatarsal I and tarsal I. The tuberosities at the bases of the metatarsals were larger than normal, presumably related to this pathology. On the cranial (dorsal) base of metatarsal III there was an erosive channel penetrating through the bone.

3.5 Discussion

Our examples show how anatomy remains a discipline of fundamental importance, for example in biomechanical analysis, and how anatomical analyis has evolved with and been enriched by new imaging technologies.

3.5.1 Velociraptor *Model*

Velociraptor is more closely related to extant birds than is *Tyrannosaurus* (e.g., Gauthier 1986; Sereno 1999), the other dinosaur for which muscle moment arm-joint angle relationships have been estimated (Hutchinson et al. 2005). Although both are Late Cretaceous terminal taxa within long-lived lineages (deinonychosaurs and tyrannosauroids, respectively); both dating back to at least the Late Jurassic (Sereno 1999; Xu et al. 2003) and have lost some ancestral features of those lineages and gained some derived features (especially larger size), they still have the potential to reveal anatomical and functional transitions within the hindlimb along the line from coelurosaurian dinosaurs to extant birds. This is particular true for those anatomical details that are retentions of features that are ancestral along this line (e.g., Gatesy 1990; Carrano 1998; Hutchinson and Gatesy 2000), despite the distance of the terminal taxa from the line itself. Hence as long as one is cautious about the relative antiquity of features in such terminal taxa (by using proper outgroup analysis and considering more/less primitive traits in related taxa that are not modelled), biomechanical models of these taxa can still reveal the evolutionary polarity of functional traits.

Many hip muscles showed relatively larger moment arms (even with size factored out) in *Tyrannosaurus*. This was likewise generally the case for distal muscle groups including the knee, ankle, and toe extensors and flexors (not shown). This matches the expectation of some positive allometric scaling of muscle moment arms from smaller to larger theropods (e.g., Alexander et al. 1981; Biewener 1989, 1990; Hutchinson 2004b; Hutchinson et al. 2005) – larger moment arms are needed to maintain similar peak stresses on the musculoskeletal system during the support of body weight or protraction of relatively heavier limb segments. However more data from other extant and extinct theropod taxa are needed to test whether this is actually allometric scaling or differences in phylogenetic position. In particular, the shortened postacetabular ilium (and reduced tail) of maniraptoran theropods (e.g., Xu et al. 2003) might explain the smaller hip extensor moment arms for most muscles originating from there.

The more unusual results of the *Velociraptor* model when compared with *Tyrannosaurus* can partly be explained by phylogenetic differences in anatomy between basal coelurosaurs (e.g., tyrannosaurs) and maniraptoran theropods (e.g., deinonychosaurs). *Velociraptor* has the shortened but somewhat more strongly retroverted ischia typical of its clade, which should have moved some muscle origins further caudally and transformed their moment arms. Our model does show some relatively increased moment arms for some but not all muscles with ischial (e.g., M. flexor tibialis internus 1 if present) origins. Likewise, shape changes of the preacetabular ilium and lesser trochanter/trochanteric crest in maniraptorans might explain the altered moment arms for the deep dorsal thigh muscles (Fig. 3.4) attaching to those surfaces (Hutchinson 2001a,b; Carrano and Hutchinson, 2002) in *Velociraptor*, relative to *Tyrannosaurus*.

Researchers have long discussed the effects of retroversion of the pubic bones in maniraptorans (and convergently in ornithischian dinosaurs) on thigh muscle function (e.g., Romer 1927; Walker 1977; Perle 1985; Hutchinson and Gatesy 2000). Rather than resulting in a dramatic transformation of the functions of the major pubic muscles Mm. puboischiofemoralis externus 1 and 2, our models show that this change only reduced (but did not eliminate as Hutchinson and Gatesy (2000) inferred, although this might have occurred deeper within the bird lineage) their hip flexor moment arms. A functional shift into hip extension (i.e., femoral retractors; Romer 1927; Charig 1972; Perle 1985), not flexion, is strongly rejected by the model. This conservation of hip flexor function results from the insertion of these muscles, which is still laterally positioned on the proximal femur (e.g., Hutchinson 2001b), and lies above, not below, the femoral head (*contra* Perle 1985), hence imparting a flexor moment arm to the muscle group. This very likely holds for ornithischian dinosaurs as well.

The ultimate explanation for retroversion of the pubes remains elusive – shifts of the body's center of mass and sizes or positions of digestive or respiratory tissues, for example, are at least consequences of this retroversion, but we remain agnostic about evolutionary causation (see also Hutchinson and Gatesy 2000).

3.5.2 Elephant Models

We were impressed to see in our elephant specimens how fine osteological details first described from visual observation of dried museum skeletons (e.g., Blair 1710; Smuts and Bezuidenhout 1993, 1994) could still be clearly observed with conventional CT scanners. This imaging approach also spares the considerable effort of thoroughly cleansing and degreasing bones, which often damages them. In the case of heavy elephant bones, this also renders sharing these anatomical data with other researchers much more simple than sharing the actual specimens. Furthermore, the detailed anatomy of the prepollex and prehallux has surely been obscured (e.g., see review in Eales 1929) by the emphasis of anatomists on describing cleaned osteological specimens, from which less ossified (and perhaps unexpected) structures such as these 'predigits' can easily be overlooked or lost. Modern imaging techniques help ensure that such strange calcifications are recognized and described *in situ*. In particular, the soft tissue context in which they are embedded can be analysed before destructive dissection.

The finding that elephant prepolluces and prehalluces have identifiable osteological correlates (e.g., Witmer 1995) of their articulations with other bones of the manus and pedes opens the possibility that the evolutionary history of these strange 'sixth digits' of elephantids can be traced through the fossil record. As these neomorphic digits are unknown in extant relatives (e.g., sirenians and hyraxes), they must have arisen within the Proboscidea. Presumably, as they support the digital cushions of subunguligrade elephant feet (Weissengruber et al. 2006), they originated in concert with the expansion of these digital cushions into the highly derived footpad structure of elephantids. Hence any information on the presence/absence of the articular surfaces in fossils would not only illuminate when the prepollex or prehallux first evolved, but also shed light on the genesis of this remarkable footpad.

We also have described numerous osteological correlates of soft tissue attachment in elephant limb bones that are useful for functional morphologists and paleontologists, and add biological context to characters useful in systematic studies. These features seem to be consistent between Asian and African elephant species, yet individual variation in many of these features remains very poorly understood. Quite a few of them are expressed at a very young age (e.g., Mariappa 1986).

How many toe bones do elephants have? There were no differences among the numbers of digits (five total in each manus and pes) in our specimens; we expect that deviations from this pattern are highly abnormal for any elephants. We found some different counts of phalanges and sesamoids in the manus and pes than sometimes reported for elephants. In the manus, authors seem to agree on a proximal sesamoid count of 1-2-2-2-2 but we found that there are two proximal sesamoids associated with digit I, albeit very small ones that may sometimes be co-ossified. We found that the manus phalangeal formula is 1-3-3-3-2 for Asian and African elephants (Smuts and Bezuidenhout 1993; Ramsay et al. 2001), but two of our seven adult Asian elephants had an extra digit I phalanx. Eales (1928, 1929) had similar findings for a fetal African elephant with a phalangeal formula of 2-3-3-3-2. Sikes (1971) indicated a manus phalangeal formula of 1-2-2-2-2 for African elephants; apparently the distal phalanges of digits II-IV were overlooked, or variability is even higher than suspected.

Pes digit I had no phalanges in adults, unlike our juvenile specimens, which match the fetal Asian elephant that Mariappa (1986) described as having one phalanx that was 'a minute cartilage' (p.27). Interestingly, Eales (1928, figs 6–8) showed two phalanges here in a fetal African elephant, although her various figures (e.g., 1928, figs 19–22) do not always show consistent phalangeal counts for other digits and her text (1929) contradicts this. It seems likely that the 'fused sesamoids' Mariappa (1986) noted for digit II were too small in his fetal specimen to differentiate; adults typically have two separate ossifications (*vide* Ramsay et al. 2001). If this is correct, then African and Asian elephants both typically should have a pedal sesamoid count of 1-2-2-2-2 (Smuts and Bezuidenhout 1994; Ramsay et al. 2001) although some individuals lack the hallucal sesamoid (none seem to have paired ones). Mariappa (1986) claimed three phalanges in digit II but only two phalanges in digit IV for a fetal Asian elephant. It is possible that Mariappa (1986, p. 26) referred to digit IV where digit V was intended; this matches his figure 1.17. We found the opposite in our main specimen, although this was variable among our other adults – a phalangeal formula of (0/1)-3-3-3-3 is common although some individuals, even large adults, have two phalanges for some of digit(s) II-V (although never digit III if other digits have three phalanges). Sikes (1971) again stated a phalangeal formula of 1-2-2-2-2 for the pes, which seems an oversimplification of the general pattern. Neuville (1935, figures 17,18) showed an unusual pes specimen with a possible fourth phalanx for digit III.

The patterns we have found for manus and pedal anatomy are from multiple individuals of a wide size and age spectrum. These generally match the observations for African elephants (Smuts and Bezuidenhout 1993, 1994), although differences in manual phalangeal count (e.g., manus digit I) probably also relate to misidentification of phalanges (as middle/ungual) due to the study of non-articulated cleaned bones, rather than the *in situ* specimens we have observed. Additionally, the emphasis on studying fetal or otherwise immature specimens (inevitably with less ossification and more ambiguous differentiation of morphology) has been misleading. We conclude that there are no salient differences in phalanx or sesamoid number between the species. Although some Asian elephants may have two phalanges in digit I, it is not clearly the predominant pattern, and the proximal sesamoids of manus digit I bear closer anatomical study. High individual variation in these traits and low sample sizes in most studies (1–2 usually) has exacerbated the confusion, and even our sample of 9–10 specimens is too small to draw statistically meaningful generalizations from.

Our use of intact cadaveric specimens rather than defleshed and disarticulated skeletal remains exemplifies the advantages of modern anatomical techniques. Soft tissue data and *in situ* positions are not lost (and are preserved digitally) even once dissections and skeletonization are done. Typical museum specimens have had all of this information scoured clean, reducing the quantity of information that specimens can provide, and potentially leading to anatomical misinterpretations.

3.6 Conclusions

We have provided examples from paleobiological and neontological perspectives on animal limb function (and evolution) to emphasize how the problem of conducting biomechanical analysis is solved from these perspectives, and how anatomical methods ranging from dissection, osteology, and functional morphology, to modern imaging techniques and computer modeling are used similarly or differently in these perspectives. We used an extinct dinosaur (*Velociraptor*) and an extant elephant (*Elephas maximus*) as examples of comparable subjects of interest to paleobiologists and neontologists. Extinct dinosaurs present mainly osteological evidence which can be used to investigate scaling trends of bone shape with size, to analyze simple functional morphology of bony articulations, or to construct complex computer models that add soft tissues to an articulated skeleton in order to more integratively analyze limb mechanics. Dissection of course is impossible; bones are the primary evidence and can themselves be fruitful. Yet the vexing problem that more complex analyses introduce more unknown parameters and hence broader, less specific quantitative results, can be frustrating regardless of the techniques available.

In contrast, such abundant data are accessible in living animals that complex biomechanical analysis becomes a different challenge: what to measure and how, vs. what assumptions must be made about unmeasured parameters? Many parameters may not be measured, simply because of time constraints or often because insufficient animals can be obtained, or invasive experimentation is impossible/undesirable (particularly the case for elephants).

Both perspectives hence share a common theme of balancing (and explicitly engaging with) the known and unknown in order to address research questions. Anatomical analysis forms the foundation of these perspectives as well. However, paleontologists generally proceed from the bones outward to interpret the once-intact whole organism, whereas neontologists typically dissect inwards from the surface of a whole animal, only later to reach the bones. Part of the wonder of modern imaging techniques is that they can reverse the directions of these pathways of inquiry – with computed tomography one can interpet skeletal mechanics in an extant animal without the conventional scalpel, or peer inside irreplaceable fossil bones without harm to them. Our examples show how biomechanical analysis is a field where anatomical inquiry, both classical and modern, is alive and well and offers exciting ways to integrate paleontological and neontological approaches.

Our two case studies also exemplify how new anatomical techniques have synergy with new and old methods in biomechanical analysis (computer simulation of dinosaurs) and phylogenetic studies (comparative anatomy of elephant species). Much as anatomy remains a crucial foundation for phylogenetics (for theropods e.g., Gauthier 1986; Sereno 1999; Xu et al. 2003; for elephants e.g., Shoshani and Tassy 1996; Shoshani 1998; Thomas et al. 2000), it is an often under-appreciated foundation of biomechanical analysis. For example, in order to figure out how a dinosaur stood or moved, we need to calculate how large of a moment each muscle-tendon unit could have generated about its hind limb joints and relate those moments to the moments of gravity, inertia, and ground reaction forces that standing or moving imposed. This is basic Newtonian mechanics, but relies on accurate anatomical data. Our model of muscle moment arms in *Velociraptor* takes one step toward this end, by representing the 3D geometry of the bones and muscle-tendon units. This also aids in resolving the problem that function often cannot reliably be inferred simply from anatomical form (e.g., Lauder 1995; Koehl 1996) by explicitly quantifying mechanical function, using anatomical geometric data.

The dinosaur model also shows how our elephant modeling will next proceed; surprisingly, dinosaur comparative osteology is much more well described than for elephant limbs, so we needed to fill in this gap first for extant elephants. Now that we have 3D characterizations of limb bone anatomy in Asian elephants, we can export our bone images into musculoskeletal modeling software and combine them with the properties of individual muscle-tendon units (e.g., estimated force-generating capacity) that we are measuring in dissections. Since we know how Asian elephant anatomy differs from African elephants (only very slightly), we can use models to address if and how these anatomical differences affect musculoskeletal mechanics.

Despite these epistemological differences, the fact remains that extant or extinct, dinosaurs and elephants are still organisms. As their anatomy is often same basic components that were just differently 'tinkered' (Jacob 1977) together during evolution, the same basic anatomical data are needed regardless of the species being studied, and analysis can proceed in much the same way, even if it must proceed more indirectly and tentatively for extinct taxa. The various disciplines of neontology and paleontology are often separated more by human perspective, history, and bias than by real biological differences. Anatomy is a crucial line of evidence that integrates them all.

Acknowledgments

We thank Silvia Salinas-Blemker for help constructing the *Velociraptor* model, in addition to the expertise of Vern Houston and colleagues at the New York University School of Medicine in laser scanning the bones. Mark Norell is thanked for access to the Mongolian specimens in his care, and trusting JRH to carry them away for scanning. Brian Garner provided expert help in smoothing and reformatting the bone files after scanning. Technical support, advice, and other input along the way from Scott Delp, Peter Loan, and F. Clay Anderson is much appreciated. This work was completed as part of a grant from the National Science Foundation in 2001. We also thank the staff of Whipsnade Wild Animal Park and other facilities for access to cadaveric elephant specimens, without which our research would be impossible. We thank Richard Prior in the Pathology Department of the Royal Veterinary College for aid during our dissections, and colleagues in the Structure and Motion Laboratory for additional assistance. Funding for the elephant research was provided in 2005 by the BBSRC for New Investigator grant number BB/C516844/1. Additional funding for CT scanning was provided by a University of London Central Research Fund grant in 2005. We appreciate technical support for CT scanning and MIMICS software from Chris Lamb, Renate Weller, Karin Jespers, and Materialize, Inc. Helpful input on this chapter was received from Chris Basu, Cyrille Delmer, Roland Frey, Victoria Herridge, and Gerald Weissengruber.

References

Alexander R McN (1989) Dynamics of dinosaurs and other extinct giants. Columbia University Press, New York

Alexander R McN, Ker RF (1990) The architecture of leg muscles. In: Winters JM, Woo SL-Y (eds) Multiple muscle systems. Springer-Verlag, New York, pp 568–577

Alexander R McN, Jayes AS, Maloiy GMO, Wathuta EM (1981) Allometry of the leg muscles of mammals. J Zool 194:539–552

Alexander R McN, Maloiy GMO, Hunter B, Jayes AS, Nturibi J (1979) Mechanical stresses in fast locomotion of buffalo (*Syncerus caffer*) and elephant (*Loxodonta africana*). J Zool 189:135–144

An KN, Takahashi K, Harrigan TP, Chao EY (1984) Determination of muscle orientations and moment arms. J Biomech Eng 106:280–283

Biewener AA (1989) Scaling body support in mammals: limb posture and muscle mechanics. Science 245:45–48

Biewener AA (1990) Biomechanics of mammalian terrestrial locomotion. Science 250:1097–1103

Blair P (1710) Osteographica elephantina. Phil Trans Roy Soc Lond 27:51–168

Carrano MT (1998) Locomotion in non-avian dinosaurs: integrating data from hindlimb kinematics, in vivo strains, and bone morphology. Paleobiology 24:450–469*l*

Carrano MT, Hutchinson JR (2002) Pelvic and hindlimb musculature of *Tyrannosaurus rex* (Dinosauria: Theropoda). J Morph 252:207–228

Charig A (1972) The evolution of the archosaur pelvis and hindlimb: an explanation in functional terms. In: Joysey KA, Kemp TS (eds) Studies in vertebrate evolution. Oliver & Boyd, Edinburgh, UK, pp 121–155

Coombs WP (1978) Theoretical aspects of cursorial adaptations in dinosaurs. Q Rev Biol 53:393–415

Csuti BA, Sargent EL, Bechert US (eds) (2001) The elephant's foot: care and prevention of foot conditions in captive Asian and African elephants. Iowa State Press, Ames

Delp SL, Hess WE, Hungerford DS, Jones LC (1999) Variation of rotation moment arms with hip flexion. J Biomech 32:493–501

Eales NB (1928) The anatomy of a foetal African Elephant, *Elephas africanus (Loxodonta africana)*. Part II. The body muscles. Trans Roy Soc Edinburgh 55:609–642

Eales NB (1929) The anatomy of a foetal African Elephant, *Elephas africanus (Loxodonta africana)*. Part III. The contents of the thorax and abdomen, and the skeleton. Trans Roy Soc Edinburgh 56:203–246

Farlow JO, Gatesy SM, Holtz TR Jr, Hutchinson JR, Robinson JM (2000) Theropod locomotion. Am Zool 40:640–663

Gambaryan PP (1974) How mammals run. John Wiley & Sons, New York

Gatesy SM (1990) Caudofemoral musculature and the evolution of theropod locomotion. Paleobiol 16:170–186

Gauthier JA (1986) Saurischian monophyly and the origin of birds. In: Padian K(ed.) The origin of birds and the evolution of flight. Mem Calif Acad Sci 8:1–55

Hutchinson JR (2001a) The evolution of pelvic osteology and soft tissues on the line to extant birds (Neornithes). Zool J Linn Soc 131:123–168

Hutchinson JR (2001b) The evolution of femoral osteology and soft tissues on the line to extant birds (Neornithes). Zool J Linn Soc 131:169–197

Hutchinson JR (2002) The evolution of hindlimb tendons and muscles on the line to crown-group birds. Comp Biochem Physiol A 133:1051–1086

Hutchinson JR (2004a) Biomechanical modeling and sensitivity analysis of bipedal running ability. I. Extant taxa. J Morph 262:421–440

Hutchinson JR (2004b) Biomechanical modeling and sensitivity analysis of bipedal running ability. II. Extinct taxa. J Morph 262:441–461

Hutchinson JR, Garcia M (2002) *Tyrannosaurus* was not a fast runner. Nature 415:1018–1021

Hutchinson JR, Gatesy SM (2000) Adductors, abductors, and the evolution of archosaur locomotion. Paleobiology 26:734–751

Hutchinson JR, Famini D, Lair R, Kram R (2003) Biomechanics: are fast-moving elephants really running? Nature 422:493–494

Hutchinson JR, Anderson FC, Blemker S, Delp SL (2005) Analysis of hindlimb muscle moment arms in *Tyrannosaurus rex* using a three-dimensional musculoskeletal computer model. Paleobiology 31:676–701

Hutchinson JR, Schwerda D, Famini D, Dale RHI, Fischer M, Kram R (2006) The locomotor kinematics of African and Asian elephants: changes with speed and size. J Exp Biol 209:3812–3827

Jacob F (1977) Evolution and tinkering. Science 196: 1161–1196

Koehl MAR (1996) When does morphology matter? Ann Rev Ecol Syst 27:501–542

Lauder GV (1995) On the inference of function from structure. In: Thomason JJ (ed.) Functional morphology in vertebrate paleontology. Cambridge University Press, Cambridge, pp 1–18

Mariappa D (1986) Anatomy and histology of the Indian elephant. Indira Publishing House, Oak Park, MI

Neuville H (1935) Sur quelques caractères anatomiques du pied des éléphants. Arch Mus Nat d'Hist Natur Paris 6e Série 13:111–183

Paul GS (1998) Limb design, function and running performance in ostrich-mimics and tyrannosaurs. Gaia 15:257–270

Payne RC, Veenman P, Wilson AM (2004) The role of the extrinsic thoracic limb muscles in equine locomotion. J Anat 206:193–204

Payne RC, Hutchinson JR, Robilliard JJ, Smith NC, Wilson AM (2005) Functional specialization of pelvic limb anatomy in horses (Equus caballus). J Anat 206:557–574

Perle A (1985) Comparative myology of the pelvic-femoral region in the bipedal dinosaurs. Paleontol J 19:105–109

Ramsay EC, Henry RW (2001) Anatomy of the elephant foot. In: Csuti BA, Sargent EL, Bechert US (eds) (2001) The Elephant's Foot: Care and Prevention of Foot Conditions in Captive Asian and African Elephants. Ames: Iowa State Press, pp 9–12.

Romer AS (1927) The pelvic musculature of ornithischian dinosaurs. Acta Zoologica 8:225–275

Roth VL (1984) How elephants grow: heterochrony and the calibration of developmental stages in some living and fossil species. J Vert Paleont 4:126–145

Sereno PC (1999) The evolution of dinosaurs. Science 284:2137–2147

Shoshani J (1998) Understanding proboscidean evolution: a formidable task. Trends Ecol Evol 13:480–487

Shoshani J, Tassy P (eds) (1996) The Proboscidea: evolution and palaeoecology of elephants and their relatives. Oxford University Press, Oxford

Sikes SK (1971) The natural history of the African elephant. Weidenfeld and Nicolson, London

Smuts MMS, Bezuidenhout AJ (1993) Osteology of the thoracic limb of the African elephant (*Loxodonta africana*). Onterstepoort J Vet Res 60:1–14

Smuts MMS, Bezuidenhout AJ (1994) Osteology of the pelvic limb of the African elephant (*Loxodonta africana*). Onterstepoort J Vet Res 61:51–66

Thomas MG, Hagelberg E, Jones HB, Yang Z, Lister AM (2000) Molecular and morphological evidence on the phylogeny of the Elephantidae. Proc Roy Soc Lond B 267:2493–2500

Walker AD (1977) Evolution of the pelvis in birds and dinosaurs. In: Andrews SM, Miles RS, Walker AD (eds) Problems in vertebrate evolution, Linn Soc Symp Ser 4. Academic Press, London, pp 319–358

Weissengruber GE, Egger GF, Hutchinson JR, Groenewald HB, Elsässer L, Famini D, Forstenpointner G (2006) The structure of the cushions in the feet of African Elephants (*Loxodonta africana*). J Anat 209:781–792

Witmer LM (1995) The extant phylogenetic bracket and the importance of reconstructing soft tissues in fossils. In: Thomason JJ (ed.) Functional morphology in vertebrate paleontology. Cambridge University Press, Cambridge, pp 19–33

Xu X, Zhou Z, Wang X, Kuang X, Zhang F, Du X (2003) Four-winged dinosaurs from China. Nature 6921:335–340

Zajac FE (1989) Muscle and tendon: properties, models, scaling, and application to biomechanics and motor control. Crit Rev Biomed Eng 17:359–411

4
Locomotion-related Femoral Trabecular Architectures in Primates – High Resolution Computed Tomographies and Their Implications for Estimations of Locomotor Preferences of Fossil Primates

Heike Scherf

4.1 Introduction – Bone as a Source of Information

Bones and teeth are often the only preserved items of extinct animals. Soft tissue remnants, stomach contents or tracks are only preserved under specific embedding and fossilization conditions. As palaeontology seeks to understand how extinct creatures appeared and existed, fossil bone provides the best source of information for reconstructions of fossil species. Additional information about the ecology of extinct animals may be gained from the embedding sediment and associated plant fossils. Since the beginning of palaeontology, the form and locomotor features of extinct animals were inferred from the external characteristics and proportions of their bones. For locomotor studies, their bone surface morphologies and proportions were compared with those of extant animals, with special attention to locomotor relevant features. This kind of comparative analyses may encounter difficulties if the fossil species practiced a unique locomotor pattern which can not be compared with locomotor patterns in extant forms (Day 1979). It has to be considered that locomotion is always performed by associated segments of the body, with each segment influencing the neighbouring segment. Therefore, reliable conclusions can only be drawn if almost complete skeletons are used in comparative analyses. However, fossil specimens are often incomplete or damaged and frequently only single bones are preserved. Consequently, it is difficult to interpret the functional context of fossil bones.

An additional method to interpret the main locomotor modes of fossils is presented here, focusing on the complex biomechanical interactions between internal osseous structures and locomotor loading conditions. Bone is capable of functional, load-directed modification, even though the mechanically optimal functionality of bone is currently not entirely understood (Huiskes 1997). During different habitual locomotor modes specific loads are applied on the limb bones, leaving a clear signature in the trabecular architecture. Rafferty (1998) noted, referring to 2D radiographs, that specialized modes of locomotion which cause relatively high and specific loading conditions on the hind limbs, like e.g. leaping, give rise to nonhomogeneous trabecular architectures in the proximal femur. This condition contrasts with homogeneous trabecular architectures which evolve under multiple moderate loading conditions, caused by unspecialized locomotor habits.

Characteristic features of the trabecular architecture in the proximal femur region of different extant and fossil primate species, related to respective locomotor modes, can be analysed and classified by high resolution 3D computed tomography images. This method provides deeper insight into the potential loading conditions of single fossil bones and, thereby, contributes to a better

Max Planck Institute for Evolutionary Anthropology, Deutscher Platz 6, D-04103 Leipzig, Germany estimation of locomotor performance of the respective species. In this study femora of five extant primate species, each of which has a specialized mode of habitual locomotion, are investigated (*Alouatta seniculus*, *Semnopithecus entellus*, *Papio hamadryas*, *Hylobates syndactylus/H. lar moloch*, *Homo sapiens*). The obtained results are used as a comparative base to analyse the femoral trabecular architecture of the two Miocene primate species *Paidopithex rhenanus* and *Pliopithecus vindobonensis* with regard to their preferred type of locomotion.

4.2 Adaptation of Bone to Functional Loads

Bone is a functional structure and alterations of the loading conditions affect its internal and external structures (Fig. 4.1). This has been common knowledge since the 19th century (Wolff 1892). Since then the biomechanical, morphological and histological relationships between bone and applied loads has been the subject of increasingly detailed analyses.

4.2.1 Nature of Loading Conditions and Other Factors which Influence the Character of Bone

It is assumed that loads which act upon the bones of the hind limbs are primarily related to locomotion. The loading conditions of the femur depend on body weight and muscle activity (Pauwels 1965, 1980; Duda 1996). Body weight, as a physical measure independent of the actual weight, is a constant quantity in the load history. However, the loads applied by the muscles change in accordance to the activity pattern of the muscles (Fig. 4.2). The hip muscles help to balance the body weight on the hind limbs and therefore to stabilize the posture of the trunk. They apply a constant load

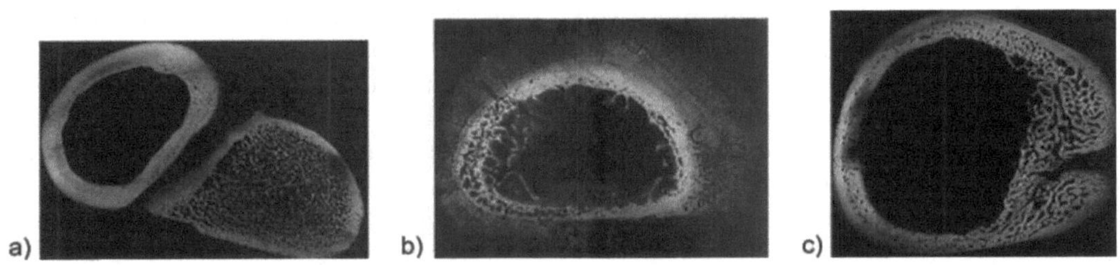

FIG. 4.1. Microradiographs of forelimb mid-shaft section of juvenile pigs. (**a**) Radius and ulna of a normal pig. (**b**) The radius undergoing an 'explosive' periosteal proliferation three weeks after removal of the ulnar diaphysis. (**c**) Three month after removal of the ulnar diaphysis the remodelling of the radius is nearly finished; altered after Goodship et al. (1979); Reprinted with permission from The Journal of Bone and Joint Surgery, Inc.

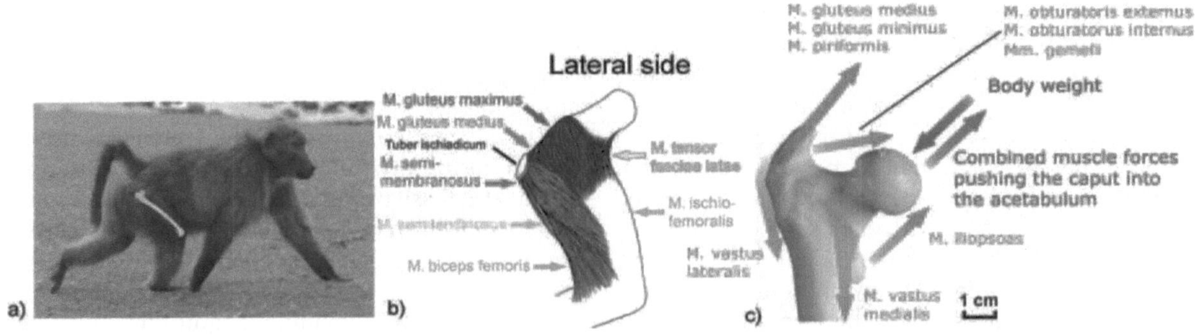

FIG. 4.2. (**a**) Walking chacma baboon (*Papio ursinus*) with outlined femur; photograph by Larissa Swedell. (**b**) Position of lateral thigh muscles which load the femur; changed after Swindler and Wood (1973). (**c**) Directions of action of proximal thigh muscles; after Scherf (2007), photograph by Anika Hebs, Senckenberg Research Institute (Germany) (see Color Plates, Fig. 4.2).

on hip and femur during various postures. In bipedal postures these muscles alone are responsible for stabilization of the body. In quadrupedal postures the muscles of the shoulder girdle additionally contribute to body stabilization. During locomotion the centre of gravity of the body is constantly changing its position relative to the joints. Therefore, body weight and stabilizing muscles apply correspondingly differing loads on the limb bones. In addition, those muscles contributing to locomotion itself apply different loads on the limb bones. These combined loads represent the main loading conditions of the hind limbs. The actual muscles and the loads they apply differ between the different types and phases of locomotion, as various muscles contract in each phase. Accordingly, the relationship between the mode of locomotion and the resulting loading conditions is very complex.

The muscle forces cause more significant effects on bone structure in male humans, whereas body weight is the more prominent factor in female humans (Welten et al. 1994). Interestingly, Hutchinson et al. noted in chapter 3 that in elephants the relatively lighter females show more pronounced muscle scars compared to the heavier males. Detailed investigations of gender differences in muscle forces in elephants could be instructive in this context, as the gender relations seems to be reversed in this group compared to humans. Heredity of body composition in terms of disposition of adipose or muscular issue is also assessed as an important factor, but is estimated to be outbalanced by lifestyle factors (Salamone et al. 1995). A mathematical model of the relationship between bone density and daily loading history of human calcaneus cancellous bone indicates that the stress intensity has a higher influence on bone density and amount of bone than the number of loading cycles (Whalen et al. 1988). Strain rate is another factor which influences bone mass. The higher the strain rate, the greater will be the effect and, therefore, the related increase of bone mass (Lanyon 1981). Trabecular bone is also influenced by the rise time of loads (Goldstein et al. 1991); the rise time of loads is the time required for a loading case to reach its maximum load. Apart from the above points, it is known that at least in humans bone reacts not only to mechanical influences but also to a variety of non-mechanical factors, comprising natural agents like hormones, vitamins, ions, amino acids, and artificial agents like hormone and vitamin analogues and drugs. However, it is assumed that their influence on skeletal physiology does not exceed that of mechanical factors (Frost et al. 1998).

4.2.2 Effect of Body Size on Bone Loading

Considering locomotor loads, the absolute body size of an individual is also of relevance. In differently sized animals the forces arising during habitual locomotion '. . . do not scale in proportion to the body weight.' (McNeill Alexander 1985, p 37). As long as it is impossible to determine the actual loading condition of and in a bone, ground reaction forces are the best obtainable measure to describe locomotor loading. The relative ground reaction forces, which refer to body weight, increase with decreasing body size for equal modes of locomotion. Therefore, smaller animals apply relatively higher forces which are several times their body weight on the ground compared to bigger animals and, vice versa, the bigger the animal, the smaller are the relative ground reaction forces (McNeill Alexander 1985; Günther 1989). Nonetheless, Biewener (1989) stated that during strenuous locomotion the muscle and bone peak stresses of animals of a large range of body sizes are fairly the same. He assumes that this is achieved by size-dependent changes in the organization of the limbs. At first sight these two statements seem contradictory, but the consistent stress level in differently sized animals can be seen as the consequence of the increasing relative ground reaction forces by decreasing body size and therefore body weight. Considering locomotor loading in certain species, Lanyon (1981) inferred from experiments with horses and dogs that their actual loading pattern remains quite the same throughout different paces and therefore different locomotor modes. Although the loading pattern remained constant, these experiments showed marked differences in the peak strains of the various locomotor regimes.

4.2.3 Bone Biology

Both bone modelling and remodelling processes act to influence the shape of bone. During modelling, bone is built and shaped. During remodelling the present bone mass is retained under consistent loading conditions. If the strain level of bone is changed by the functional loads acting on bone, remodelling adjusts the bone to the new conditions (Lanyon 1981). Increased loading, e.g., will cause bone formation (Chambers et al. 1993). The adjustment of bone structures to varying loads is likely to be the main reason for remodelling. Remodelling is also responsible for the repair of fatigue damages and the maintenance of serum mineral metabolism (Ott 1996; Huiskes et al. 2000). The remodelling process is based upon directed apposition and resorption of bone (Ott 1996). It is governed by the reciprocal activity of osteoclasts and osteoblasts combined in basic multicellular units (BMUs) or bone remodelling units (BRUs) (Dempster 1992). For adult humans the annual turnover rate of cortical bone is about 3%, whereas approximately 25% of the trabecular bone is affected (Parfitt 1983).

Cortical and cancellous bone respond in different ways to alterations of the applied loads. Cortical bone, which has the highest load capacity, reacts to differing loading conditions by a variation of its cross sectional geometry and its density distribution in the cross section (Pauwels 1965, 1980; Lanyon 1974; Adler 1998). However, Lanyon (1981) stated that 'Local cortical thickness is not proportional to local functional load . . .' (p 321). Different cortical regions are therefore subjected to different tensile or compressive functional strains and strain magnitudes. Lanyon (1981) hypothesized that genetic and mechanical factors may interact and give rise to uneven strain distribution.

Functional adaptation in cancellous bone causes alignment of trabeculae along the principle stress trajectories, known as 'Wolff's law' (Wolff 1892; Kummer 1959; Pauwels 1965, 1980). Weaver and Chalmers (1966) were among the first who found that bone strength may be influenced by the spatial organization of the trabeculae in addition to the mineral content. Evidence supporting the stress-related adaptation of cancellous bone was generated by in vivo experiments (Lanyon 1974; Goldstein et al. 1991; Guldberg and Hollister 1995; Guldberg et al. 1997a,b) and computational studies (Mullender et al. 1994; Mullender and Huiskes 1995; Huiskes et al. 2000; Tsubota et al. 2002). Finite element method (FEM) simulations of changes in trabecular architecture of the human proximal femur under single and multiple loading conditions showed a clear relation between architectural changes and apparent principal stresses in different regions of the femur (Tsubota et al. 2002). Goulet et al. (1994) assumed on the basis of structural and mechanical investigations of cancellous bone a relationship between the loading environment and the spatial organization of cancellous bone as well as the shape of the individual trabeculae. They further proposed that mechanical loading has a direct influence on the trabecular thickness while the numbers of trabeculae are affected by hormones or chemical substances.

The scanning electron microscopy (SEM) study of Whitehouse and Dyson (1974) demonstrated the variability of the trabecular architecture throughout the proximal part of the femur, which is likely to be associated with different loading conditions in the different bone regions. Recent studies verified this finding in proximal human tibiae (Williams and Lewis 1982) and in the greater trochanter and neck region of human femora (Morgan and Keaveny 2001).

4.3 High Resolution Computed Tomography (CT)

Whitehouse et al. (1971) and Whitehouse and Dyson (1974) have noted the limited significance of studies depicting the spatial trabecular architecture in two-dimensions, such as provided by thin sectioning. They tried to minimize this limitation by the use of SEM. A recently developed alternative to difficult SEM studies and time-consuming, labour-intensive, and destructive sectioning techniques is high resolution CT. High resolution CT provides detailed three-dimensional imaging of the spatial structure in cancellous bone, which is indispensable for detailed analysis of the trabecular architecture (Compston 1994; Borah et al. 2001). The analysis of the trabecular architecture with this method has already been applied in medical sciences (Goulet et al. 1994; Guldberg et al. 1997a; Müller et al. 1998; Stenström et al. 2000; Borah et al. 2001; van der Linden et al. 2001) and recently in anthropological studies (Fajardo and Müller 2001; MacLatchy and Müller 2002; Ryan and Ketcham 2002a,b). As high resolution CT is non-destructive, the material remains intact and can be used for future investigations, which is particularly important for fossil specimens.

The above-mentioned studies were carried out with medical biopsy micro-computed tomography systems, which are optimized for extant bone material. However, fossilized material requires special CT imaging conditions which are not matched by these systems (Scherf et al. 2005). During the fossilization process, bone material is altered by exchange processes, and the medullary cavity, including the space between the trabeculae, is often filled in by some mineral substance. These fillings can be composed of different minerals, depending on the surrounding sediment and the ion content of the pore water (Martill 1991). Both conditions, the alterations of the former bone substance and the fillings, cause a higher X-ray absorption rate of the specimen and complicate CT imaging techniques.

In order to account for the unique conditions of fossilization, the samples used here were imaged with a special CT system. The high resolution computed tomography system (RayScan 200, Hans Wällischmiller GmbH, Gemany) at the Foundry Technology centre of the Aalen University of Applied Sciences in Southern Germany matched the requirements to yield images of the trabecular architecture in fossil bones which were processible for histomorphometric analyses (Fig. 4.3). This system was designed for material testing of a wide range of objects and materials. The microfocus X-ray tube of the RayScan 200 is infinitely variable up to 225 kV and provides a cone-beam X-ray. This system permits imaging of samples up to a maximum diameter of 600 mm (Simon et al. 2001). The CT system is equipped with an amorphous silicon area detector with 1024 × 1024 elements and can achieve a nominal resolution of 5 µm (personal communication M. Simon, Hans Wällischmiller GmbH).

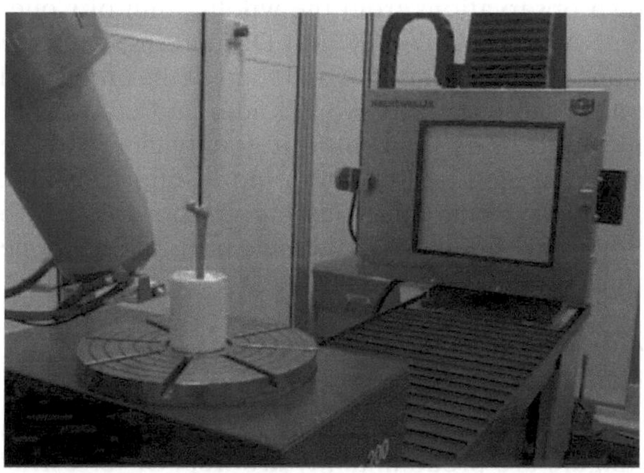

FIG. 4.3. High resolution CT System Rayscan 200 at the Foundry Technology centre of the Aalen University of Applied Sciences; Scherf (2007) (see Color Plates, Fig. 4.3).

Prior to imaging, the femora were mounted on a turntable with the shaft axis aligned along the z-axis of the imaged volume. The proximal part of the femur, from head to end of lesser trochanter was imaged in each specimen. The resolutions of the CT images differ between the samples in accordance to the size of the specimen. When investigating species of different body sizes, it must be taken into consideration that trabecular dimensions may scale with body size as a consequence of the mechanical adaptation of cancellous bone to functional loads, generated by body weight and muscles. Although the stresses in bones stay quite the same (Biewener 1989), the dimension of the bones change with changing body size and in this way the actual loads have to change, too, to maintain similar stress levels. The loads applied to the bone are expected to decrease correspondingly with the decrease in body size. Therefore, it is considered indispensable to image the specimens with the highest obtainable resolution in order to image the cancellous structure accurately. Any intentional limitation of resolution in smaller specimens which would enable the matching of resolution in bigger specimens would yield inadequate results. Due to the partial volume effect, some structures would appear bigger than actual and very fine structures would be deleted in the CT images. The partial volume effect has severe effects on structures imaged with only a few pixels across their diameter. For example, at the boundary of a structure, where two materials with two different grey values are present, an imaging pixel with a medium grey value will be created and assigned to one of these materials. Thus, the volume of one material will be enlarged by a fraction actually belonging to the other material, while the volume of the other material will be decreased or even deleted.

4.4 Investigation of Proximal Femoral Cancellous Bone of Extant and Fossil Primates

4.4.1 Investigated Species

The extant sample comprises the five genera *Alouatta seniculus*, *Semnopithecus entellus*, *Papio*

TABLE 4.1. Extant sample.

Species	*n*	Collection number	Institute
Alouatta seniculus	2	25545 ♀, 25544 ♀	Senckenberg Research Institute
	1	69.19 ♂	Dr. Senckenbergische Anatomie
Semnopithecus entellus	3	4734, 4745, 4746	Institute of Zoology (Hamburg University)
Papio hamadryas	1	1.553 ♀	Senckenberg Research Institute
	2	HaVIII83 ♂, HaVIII3 ♂	Dr. Senckenbergische Anatomie
	1	3212 ♂	Phyletic Museum Jena
Hylobates syndactylus	2	6983 ♂, 52.36 ♀	Dr. Senckenbergische Anatomie
Hylobates lar moloch	1	47979 ♂	Senckenberg Research Institute
Homo sapiens	4	10 ♀, 11 ♂, 21 ♂, 22 ♀	Senckenberg Research Institute

hamadryas, *Hylobates syndactylus*, *H. lar moloch,* and *Homo sapiens*. They were provided by the Senckenberg Research Institute (Frankfurt/Main, Germany), the Dr. Senckenbergische Anatomie (Johann Wolfgang Goethe-University, Frankfurt/Main, Germany), the Phyletic Museum Jena (Germany), and the Institute of Zoology of the University of Hamburg (Germany) (Table 4.1). These genera were chosen by virtue of their preferred mode of locomotion, with the main premise that each species exhibit a specialized mode of habitual locomotion, differing from those of the other species. Due to the load-directed alignment of cancellous bone, it is assumed that differences in locomotion leave clear signatures in the trabecular architecture.

Concerning the non-human primates, the influence of body weight and size effects was also considered in the selection procedure. Therefore, differences in body weight and size were kept as small as possible under the given conditions. To exclude influences of unnatural locomotion, or altered nutrition, as well as ontogenetic effects, only femora of wild shot and adult individuals have been used, except *Semnopithecus entellus* of which only sub-adult specimens of wild shot individuals were available. In the following section the locomotor behaviour of the extant species is outlined.

4.4.1.1 *Alouatta seniculus* – Red Howler Monkey

Howler monkeys spend most of their time in the treetops and are rarely seen on the ground. They spend up to 80% of their active daytime resting and travel only short distances (Welker and Schäfer-Witt 1988). Observations on the different species of howler monkeys have yielded similar results. Therefore, it was considered to be acceptable to refer to these observations for the description of *Alouatta seniculus*.

Compared to other primates, howler monkeys climb and walk slowly and carefully, using their prehensile tail like an extra grasping limb to secure themselves during movement. The body is predominantly held in a pronograde position, i.e. the long axis of the body is parallel to the support. Leaping is rarely observed and they prefer to climb from one tree to another. If required, they can perform jumps over 3–4 m. To accomplish a jump they push off the body with their hind limbs, while one or both hands initially leave the substrate. During the jump the hands reach out to seize a branch at the final destination. Finally the feet and the prehensile tail loose the grip on the starting branch. The main push off force seems to come from the hind limbs with only a little help by the fore limbs. Even though howler monkeys move slowly most of the time, they can retire/escape quickly in situations of danger (Krieg 1928; Carpenter 1934; Welker and Schäfer-Witt 1988; own observations from the wildlife film of Collet and Vienne 1986–1989a). During quadrupedal stance and gait the fore and hind limbs are placed under the body axis, while knees and elbows are flexed to bring the center of gravity closer to the substrate and by this into a more stable position. During walking and climbing no more than two limbs are lifted from the substrate at the same time (Grand 1968a).

4.4.1.2 *Semnopithecus entellus* – Hanuman Langur

Depending on their environment Hanuman langurs are predominantly arboreal or terrestrial (Vogel and Winkler 1988). It seems reasonable to expect differences in the trabecular architecture between predominantly arboreal and predominantly terrestrial individuals of this species owing to differences in hind limb loading during arboreal and terrestrial locomotion. The locality data indicate that the individuals used herein have lived in a subtropical forest and, thus, are considered arboreal.

Regarding the locomotor and postural behaviour of predominantly arboreal *Semnopithecus entellus*, Nikolei (2002) noted that their locomotor activities take up 3.5% of their active day hours on average, while they spend over 90% resting in a sitting posture. They are frequently descending to the ground and their locomotor activities comprise equal shares of arboreal and terrestrial locomotion. Arboreal and terrestrial locomotion involve quadrupedal walking, trotting, and galloping, whereas walking is the most frequent locomotor mode. Walking and trotting is accomplished by symmetric lateral limb movements. Asymmetric limb coordination can be observed during galloping, including real flight phases. A further preferred locomotor mode is leaping. The propulsive power for the leaps is mainly provided by the hind limbs. Leaps up to 10 m between trees have been observed (Fig. 4.4). The flexibility of the substrate is often used as a spring board to enhance performance. To reduce shock on the joints during touch down, the impulse force of the leap is transformed into motion force for quadrupedal walking, as far as the environment permits such a transformation. Along with leaping, another load intensive locomotor activity of *Semnopithecus entellus* consists of dropping down from a branch after hanging from one or both forearms. A further common kind of locomotion is climbing. Vertical substrate climbing is accomplished by pushing the body upwards with the hind limbs while the fore limbs grasp the substrate. Thinner branches are climbed in a walking mode with diagonal limb coordination. Hanuman langurs also perform bipedal walking occasionally. This mode involves the legs being spread wide apart with the arms sometimes used to balance the body.

4.4.1.3 *Papio hamadryas* – Sacred Baboon

The following descriptions are based on the observations of Hall (1962) on Chacma baboons (*Papio ursinus*) from southern Africa and observations made from two wildlife films on yellow baboons (*Papio cynocephalus*) (Borland 1991; Collet and Vienne 1986–1989b). As all baboons are quite similar in their locomotor performances and as yellow and sacred baboons inhabit comparable habitats, this substitution is estimated to be suitable.

Most of their time baboons spend on the ground. Usually, they use a quadrupedal walk in terrestrial and arboreal locomotion. During walking the limbs are coordinated diagonally. The more powerful locomotor modes of galloping and trotting are only performed on the ground (Fig. 4.5). They are accomplished by asymmetric limb coordination. The time difference between the left and the right fore and hind limbs hitting the ground decreases with increasing speed. During climbing, only the hands clutch the substrate while the feet are placed flat. Baboons are observed to climb vertically with the hind limbs pushing the body upwards by successional leaps, while the hands seize the log. Shorter distances on vertical substrates are climbed in a form of pace. However, on almost vertical slopes Hall (1962) described a

FIG. 4.4. Leaping Hanuman langur (*Semnopithecus entellus*); photograph by Paul Winkler, Georg-August-Universität Göttingen (Germany) (see Color Plates, Fig. 4.4).

FIG. 4.5. Running adult female chacma baboon (*Papio ursinus*) with infant on belly; photograph by Larissa Swedell, City University of New York (USA) (see Color Plates, Fig. 4.5).

FIG. 4.6. Slowly brachiating gibbon (*Hylobates lar*) in Khao Yai National Park, Thailand; photograph by Claudia Barelli, Max Planck Institute for Evolutionary Anthropology (Germany) (see Color Plates, Fig. 4.6).

human-like climbing fashion which is accomplished by finding hand-holds first and pulling up the body afterwards. When descending steep cliffs or rocks, baboons might climb sideward and secure themselves with their hands which are kept above the level of their feet. The rare leaps are rather short and performed carefully. The forearms are put forward while leaping to seize a first hold on the other side. In yellow baboons it was sometimes observed that they let themselves drop from a higher towards a lower branch. During the fall they put the arms downward to grip the lower substrate.

4.4.1.4 *Hylobates syndactylus*/*H. lar moloch* – Siamang/Silvery Gibbon

Due to the limited number of available specimens meeting the requirements mentioned above, specimens from two gibbon species were examined. However, as the locomotor modes of all gibbons are quite similar, the results are not expected to be markedly affected.

Gibbons live arboreally and brachiation is their predominant mode of locomotion (Fig. 4.6). Brachiating gibbons are able to 'fly' across distances of up to 10 m (Preuschoft 1988). They tuck up their legs during brachiation, especially when they move fast (Carpenter 1940). The fast, ricochetal kind of brachiation is more frequently practiced by smaller gibbon species and juvenile Siamang than by adult Siamang (Fleagle 1976). Downward leaps at the end of a normal swinging phase or starting out of a resting position are also part of the locomotor repertoire of gibbons (Carpenter 1940; Fleagle 1976). The arms seem to provide the main take-off force by pulling the body forward. In this way gibbons may leap over a horizontal distance of up to 10 m, while they can cover more than 20 m in vertical distance. Apart

TABLE 4.2. Fossil specimens.

Species	n	Collection number	Institute
Pliopithecus vindobonensis	3	O.E. 304, O.E. 559, O.E. 560	Natural History Museum Basel
	3	1970/1397/22, 1970/1397/23, 1970/1398/2	Natural History Museum Vienna
Paidopithex rhenanus	1	Din 45	Hessian State Museum Darmstadt

from brachiation, climbing is the most common mode of locomotion in which the fore limbs also seem to dominate. Climbing is more often observed in Siamangs than in smaller gibbons (Fleagle 1980).

The hind limbs of gibbons are loaded during bipedal walking which is performed on the ground as well as in trees. During this locomotor mode the hind limbs are flexed at the knee and hip joints. The arms may be stretched out sideways above the level of the head like a balancing pole, or are put to the ground in a knucklewalking fashion while the body is in an partially upright posture. Unlike human walking, bipedal locomotion of gibbons also contains bouncing elements. Gibbons use bipedal locomotion more often than all other non-human primates: 5 to 12% of their active time (Carpenter 1940; Fleagle 1976; Napier 1976; Preuschoft 1988; Vereecke 2006).

4.4.1.5 *Homo sapiens* – Human

Bipedalism is just one example of the different modes of habitual locomotion investigated here. Only the basics about human locomotion are given. It has to be noted that humans mostly walk and are therefore assigned as bipedal walkers (Fig. 4.7). Running or climbing is performed to a lesser extent in daily life, depending on the individual. The preferred postures are standing, sitting and lying. More variations in human locomotion and posture surely exist but the above named habits are the most common and in this context the only relevant ones.

3D high resolution CT images of the femora of the Miocene species *Pliopithecus vindobonensis* and *Paidopithex rhenanus* are compared to those of the above listed extant species. The femur of *Paidopithex rhenanus* was placed at my disposal by the Hessian State Museum Darmstadt (Germany). Three femora of *Pliopithecus vindobonensis* were provided by the Natural History Museum Basel (Switzerland) and three more by the Natural History Museum Vienna (Austria) (Table 4.2).

4.4.1.6 *Pliopithecus vindobonensis*

The *Pliopithecus vindobonensis* specimens were found in a fissure near Neudorf an der March (Děvínská Nová Ves, Slovakia). According to their stratigraphic position they occurred in the transition of the Mammal Neogene Zone MN5 to MN6 (Daxner-Höck 1998) and are approximately 15 Million years old. The locomotion of *Pliopithecus*

FIG. 4.7. Walking humans; © Max Planck Institute for Evolutionary Anthropology (Germany) (see Color Plates, Fig. 4.7).

vindobonensis has been under debate since it was first described in detail by Zapfe (1960). Zapfe concluded in 1960 that *Pliopithecus vindobonensis* was a locomotor generalist, which did not live strictly arboreally but also spent some time on the ground. However, subsequently it was agreed that this species lived arboreally (Simons and Fleagle 1973; Szalay and Delson 1979; Fleagle 1983; Langdon 1986; Rose 1994). In the latter cited studies *Pliopithecus vindobonensis* was compared to howler monkeys (*Alouatta*) and spider monkeys (*Ateles*), which are both semibrachiators. However, the importance of suspensory locomotion and the practice of load intensive locomotor behaviours like quadrupedal running or leaping remain under debate. Currently, a final answer to the preferred locomotion of *Pliopithecus vindobonensis* cannot be given.

4.4.1.7 *Paidopithex rhenanus*

The specimen of *Paidopithex rhenanus* is also known as 'Eppelsheim femur' (Eppelsheimer Femur). It was the first fossil part of a hominoid ever found (Fig. 4.8). This single primate specimen was discovered 1820 in a sand pit in the Mainz Basin (Germany) and is approximately 9.5 Million years old (Franzen et al. 2003). Since its discovery, *Paidopithex rhenanus* caused numerous scientific debates on its taxonomic relations along with its preferred mode of locomotion. In the following, only the latest studies are mentioned.

A morphometric analysis by McHenry and Corruccini (1976) identified similarities between *Paidopithex rhenanus*, *Pliopithecus* and *Hylobates*, and recommended to assign it together with *Pliopithecus* into the subfamily Hylobatinae. Szalay and Delson (1979) classified *Paidopithex rhenanus* as a synonym of the genus *Dryopithecus* and thus as a member of the subfamily Ponginae.

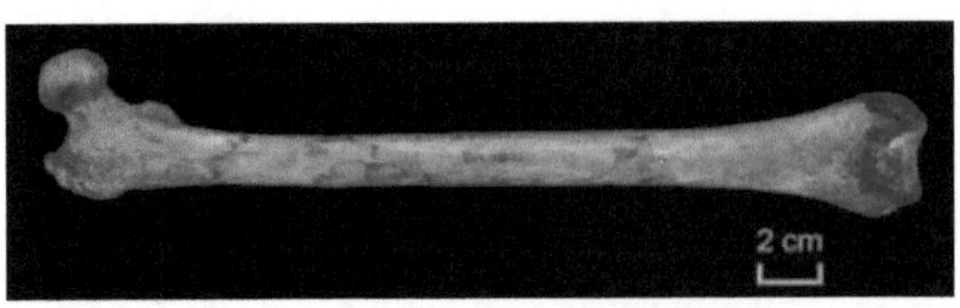

FIG. 4.8. Ventral view of the right femur of *Paidopithex rhenanus* (Din45); after Scherf (2007), photograph by Anika Hebs, Senckenberg Research Institute (Germany) (see Color Plates, Fig. 4.8).

Begun (1992) concluded from morphological analyses that the Eppelsheim femur is rather similar to atelines and *Pliopithecus* than to hominoids. Therefore, he estimated that *Paidopithex rhenanus* should belong to a separate subfamily, the Pliopithecinae. In a functional interpretation he proposed that *Paidopithex rhenanus* could have practiced agile and speedy movements and concluded that this fossil species might have been habitually hanging on its hind limbs. Due to its morphological similarities with *Pliopithecus*, Begun (1992) interpreted that *Paidopithex rhenanus* exhibited a postural behaviour comparable with *Alouatta*. Andrews et al. (1996) rejected Begun's opinion and assigned the Eppelsheim femur to the Dryopithecinae and thus to the Hominoidea, exclusive of gibbons.

4.4.2 Locomotion-related Characteristics of Cancellous Bone

The arrangement of cancellous bone reflects the different loading conditions caused by different kinds of locomotion. As in external bone morphology, every species possesses its characteristic cancellous bone architecture. The accentuation of trajectories or fainter reinforcement structures varies between individuals. In larger individuals, i.e. mainly in males, these structures are more pronounced than in smaller and female individuals. This goes along with the trabeculae tending to be thicker in large individuals of a species but definitely thinner in small individuals. Further, there seems to be a relation between the shape of the trabeculae and the body size of the individuals in some species, e.g. in *Papio hamadryas*, *Semnopithecus entellus*, and in *Homo sapiens*. Larger individuals have a higher amount of plate-like trabeculae while smaller individuals possess more rod-like or prolate trabeculae. The difference between these two types of trabecular shape is that prolate trabeculae have an elliptic cross section, while a rod-like trabeculae have a round cross section. However, in *Alouatta seniculus* and *Hylobates syndactylus/H. lar moloch* no definite relation between trabecular shape and individual body size is recognizable.

It was hypothesized in Scherf et al. (2005) that plate-like trabeculae can transfer higher loads from various directions in the plane of their extension than from directions perpendicular to the plane of their extension. In contrast, rod-like trabeculae can

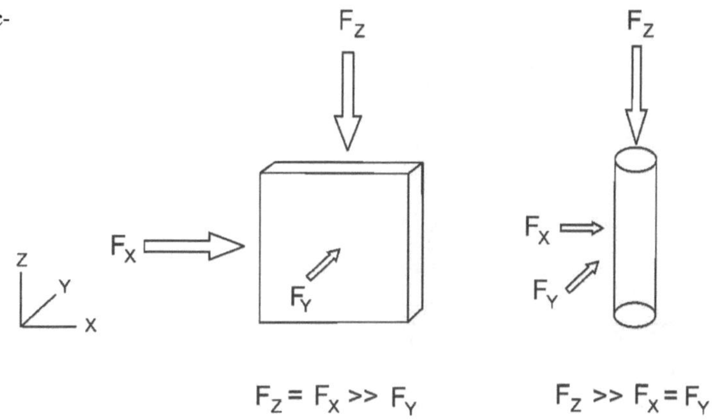

FIG. 4.9. Differences in load transmitting capacity of rod-like and plate-like structures.

transfer high loads only when they are loaded axially, i.e. in the direction of their, main extension, and just relatively low loads can be transmitted from the two remaining spatial directions (Fig. 4.9). This hypothesis is now being extended to include the influence of body size on trabecular shape. As large individuals exert higher loads on their femora than small individuals, the loads acting on similar sized trabeculae would be higher in larger individuals than in smaller ones. An increase of bone mass by formation of plate-like trabeculae does not only adjust the bone mass to the increased loading condition but also provides an extra safety factor in permitting the transfer of loads from other directions beside the direction of the main extension of a rod-like structure. By forming a plate-like structure a reinforcement in that directions in which loading occurs second most is created, according to the enhanced abilities of plate-like structures to transmit relative higher loads (Fig. 4.9). The relation between body size and trabecular shape was found only in those species in which habitual locomotor modes included intensive hind limb loading. This might require a more robust structure of the trabeculae whereas trabecular structures in *Alouatta seniculus* and *Hylobates*, which rarely apply high loads to their hind limbs, can remain more delicate.

In the following, the snap-shots of the 3D CT images (Figs. 4.10–17) of cancellous bone were all taken from femora in comparable positions. The virtual section plane always runs through the centre of the femoral head and the greater trochanter. These images were obtained with the software system VGStudio MAX (Volumegraphics, Germany).

4.4.2.1 *Alouatta seniculus* – Red Howler Monkey

The relatively low loads subjected to the hind limbs of *Alouatta seniculus* during its preferred climbing locomotion are reflected by an open cancellous bone structure with two weak preferred directions of alignment in the region of the femoral head and neck (Fig. 4.10). During climbing the loading direction varies in accordance to the position of the limb which adjusts to different substrates. The variability of limb postures is highest during climbing, compared to other kinds of locomotion, during which the thigh gets moved mainly in para-sagittal plains. Therefore, no definite trajectories evolve by predominant climbing movements. High loading conditions which occur in jumping occur very rarely and thus do not cause an adjustment of cancellous bone, such as reinforcement structures or trajectories.

Referring to the investigation of Grand (1968b) the Musculus gluteus maximus, M. gluteus minimus, and M. gluteus medius, fused with the M. piriformis, insert at the greater trochanter of the howler monkey. The M. gluteus maximus was identified as extensor, while M. gluteus minimus, M. gluteus medius and M. piriformis were described as abductors and rotators. All these muscles are being strained to a lesser extent during climbing compared to a locomotor mode involving frequent jumping and running, thus they do not need to generate high propulsive forces. In accordance with the relative low loads acting on it, the greater trochanter of *Alouatta seniculus* is small and supported by relatively few trabeculae. The mainly plate-like trabeculae of the greater trochanter are

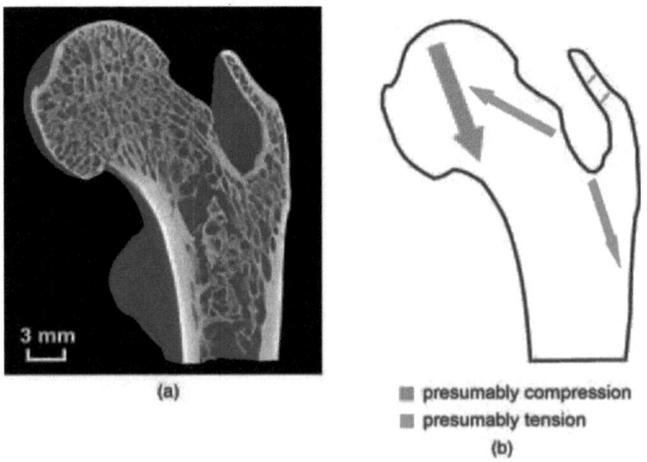

FIG. 4.10. (**a**) Ventral view of the trabecular architecture of the left femur of *Alouatta seniculus* (Senckenberg Research Institute, 25544 ♀), Scherf (2007), (**b**) Outline of trabecular alignment (see Color Plates, Fig. 4.10).

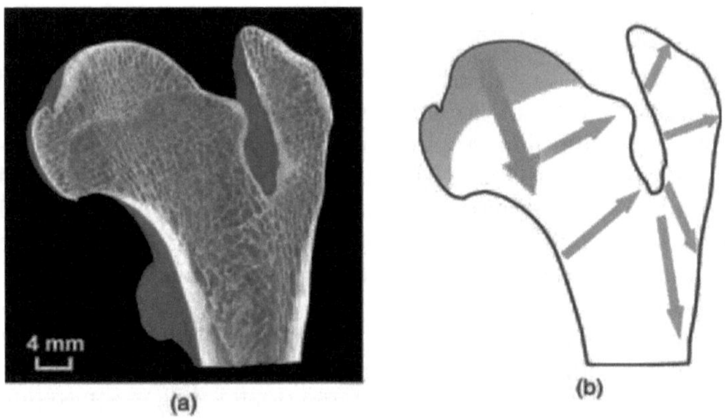

FIG. 4.11. (**a**) Ventral view of the trabecular architecture of the left femur of *Semnopithecus entellus* (Institute of Zoology of the University of Hamburg, 4734, sex was not registered) most mature specimen, Scherf (2007), (**b**) Outline of trabecular alignment (see Color Plates, Fig. 4.11).

extending perpendicular to the cortex. The plate-like shape of these few trabeculae can be seen rather as a result of varying loading directions than as a sign of high loading. The trabecular alignment in the lesser trochanter of *Alouatta seniculus* changes from an alignment in the medisagittal plain anteriorly to an alignment in a sub-transverse plane posteriorly. This seems to be a consequence of the different directions of action of the flexor muscle of the thigh (M. psoas major; Grand 1968b) probably induced by climbing movements.

4.4.2.2 *Semnopithecus entellus* – Hanuman Langur

Compared to *Alouatta seniculus*, the agile and rather load-intensive locomotion of *Semnopithecus entellus* causes a denser and more intensively connected trabecular network (Fig. 4.11). A strong trajectorial bundle runs from the lower part of the femoral neck through the femoral head to its proximal end. This trajectory can transmit high loads which act directly on the femoral head. These loads occur during running and leaping, when the femoral head gets pressed into the acetabulum by contracting muscles and body weight. Above the epiphyseal split, which is clearly visible in two of these three sub-adult individuals, a generally condensed cancellous structure is present in the femoral head, resembling a cap-like structure. As all specimens are immature, a definite statement about this feature cannot be given, as it could be linked with ontogenetic development as well as with locomotor loading.

The robust shape of the greater trochanter, its higher position relative to the femoral head, and its complex cancellous bone structure are likely to be caused by high stresses brought about by the muscles inserting on the greater trochanter. Among these muscles, the extensor muscles of the thigh generate the main propulsive force during running and jumping, and when climbing a vertical substrate. The trabecular architecture of the lesser trochanter was clearly visible only in the most mature specimen, while the sub-adult individuals showed dense and less directed trabecular architectures, disturbed by the growth plate. Therefore, concerning the legger trochanter, no reliable conclusion can be drawn. However, as the trabeculae in the most mature specimen are all inclined laterally, a strong flexor muscle force acting in medial direction on the lesser trochanter can be assumed.

4.4.2.3 *Papio hamadryas* – Sacred Baboon

Papio hamadryas can be characterized as a quadrupedal, terrestrial primate. Especially the heavy males put high loads on their hind limbs while running. The trabecular architecture displayed by the 3D CT images shows a close network of relatively thick trabeculae in the superior part of the femoral head (Fig. 4.12). This structure is somewhat similar to the upper part of the well-known 'compressive trajectory' in *Homo sapiens* and can be interpreted in a similar way. In humans the main loads of bipedal locomotion act on the superior region of the femoral head across an angle of approximately 20°, corresponding to the 'compressive trajectory'. In comparison to bipedal walking, the body of *Papio hamadryas* has not to be balanced on two limbs only and, thus, it is possible to rotate the femoral head around a larger angle during fast quadrupedal locomotion. In this way high loads are transmitted across a larger area of the femoral head than during bipedal locomotion. Accordingly, this close network might be a sign of the high loads applied during quadrupedal running. A similar feature was found in *Semnopithecus entellus*, but due to the sub-adult age of the sampled individuals and the close distance to the epiphyseal split, its relation to locomotor loading remains unclear.

The greater trochanter of *Papio hamadryas* is composed of a complex trabecular architecture and overtops the femoral head. It is comparable to the greater trochanter of *Semnopithecus entellus* and can be interpreted in the same way: the shape and the trabecular architecture of the greater trochanter in *Papio hamadryas* are potentially reflections of strong extensor muscles, contributing to the propulsive force during fast quadrupedal locomotion. The M. gluteus medius, M. gluteus minimus, and M. piriformis are inserting into the greater trochanter of the drill, *Papio leucophaeus*, and are involved in extension, abduction, inward rotation, and outward rotation (Fischer 1961). Similar functions are assumed for *Papio hamadryas*. In the lesser trochanter of *Papio hamadryas* two directions of trabecular alignment are noticeable. In the anterior half the trabeculae are inclined laterally while in the posterior-most region the trabeculae are aligned in a sub-transverse plane. Between these two areas a fine trabecular network without a definite direction of alignment is present. The different directions of alignment may reflect different directions of action of the flexor muscle inserting on the lesser trochanter (M. iliopsoas, according to observations in *Papio leucophaeus*; Fischer (1961). The different gaits may be the reason for the change in the directions of action of the flexor muscle in *Papio hamadryas*.

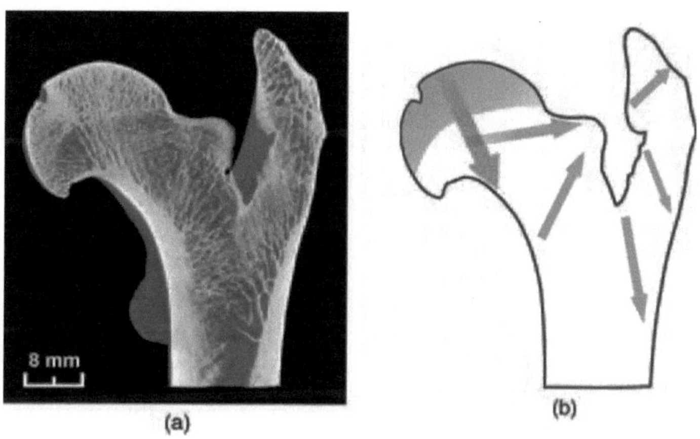

FIG. 4.12. (**a**) Ventral view of the trabecular architecture of the left femur of *Papio hamadryas* (Dr. Senckenbergische Anatomie, Ha VIII 3 ♂), Scherf (2007), (**b**) Outline of trabecular alignment (see Color Plates, Fig. 4.12).

4.4.2.4 *Hylobates syndactylus/H. lar moloch* – Siamang/Silvery Gibbon

The gibbons, *Hylobates syndactylus* and *Hylobates lar moloch*, have a dispersed trabecular bone architecture (Figs 4.13, 4.14). The distribution of cancellous bone is almost homogeneous. The main difference between the two gibbon species consists in the spatial density of the trabeculae. *Hylobates syndactylus* has a higher spatial density of trabeculae compared to *Hylobates lar moloch*. This might be a consequence of the overall higher loads put on the femora of *Hylobates syndactylus* by its higher body weight. Both species show two sheaves of plate-like trabeculae, running from the femoral neck into the head of the femur. In *Hylobates syndactylus* the inferior sheaf crosses the femoral head and ends at the cortex above the fovea capitis, while in the silvery gibbon both sheaves can only be tracked to the middle of the femoral head. The accentuation of the inferior sheaf in *Hylobates syndactylus* is probably related to the increased practice of climbing, i.e. putting about frequent loads on the hind limbs, whereas *Hylobates lar moloch* climbs less frequently. The greater trochanter is quite small in all gibbon specimens but definitely larger compared to *Alouatta seniculus*. Its architecture resembles that of the greater trochanter in *Papio hamadryas*. As the hind limbs are tucked up during brachiation, the abductor muscles inserting into the greater trochanter will experience stress more frequently and at a higher level in *Hylobates* compared to the climbing loads of *Alouatta seniculus*. However, this stress is certainly lower in the gibbon's greater trochanter compared to that of baboons. Accordingly, the greater trochanter of *Hylobates syndactylus/H. lar moloch* is definitely smaller compared to *Papio hamadryas*.

The cancellous bone network of the lesser trochanter shows different directions of alignment. In both gibbon species the trabeculae of the anterior

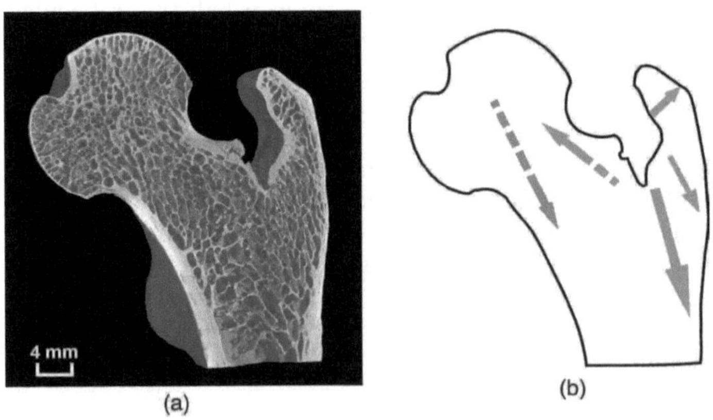

Fig. 4.13. (**a**) Ventral view of the trabecular architecture of the left femur of *Hylobates syndactylus* (Dr. Senckenbergische Anatomie, 52.36. ♀), Scherf (2007), (**b**) Outline of trabecular alignment (see Color Plates, Fig. 4.13).

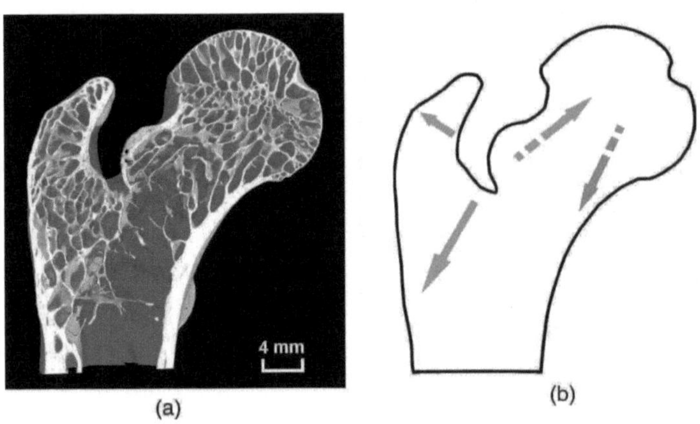

Fig. 4.14. (**a**) Ventral view of the trabecular architecture of the right femur of *Hylobates lar moloch* (Senckenberg Research Institute, 47 979 ♂), Scherf (2007), (**b**) Outline of trabecular alignment (see Color Plates, Fig. 4.14).

half are inclined laterally, while in the posterior half *Hylobates lar moloch* shows an undirected network and in the *Hylobates syndactylus* specimens the trabeculae are reoriented and aligned in transverse direction. A similar change of alignment as in *Hylobates syndactylus* was found in *Alouatta seniculus*. Correspondingly, it can be interpreted, as the result of a change in the direction of action of the flexor muscles inserting on the lesser trochanter. Differing muscle strains during climbing, walking, and tucking up the legs during brachiation of *Hylobates syndactylus* might be responsible for such a change in direction. *Hylobates lar moloch* climbs less and brachiates more compared to *H. syndactylus*, probably straining its flexor muscles in a more uniform way which does not give rise to a second direction of trabecular alignment in the lesser trochanter. However, considering the small sample size, the inferred differences between the two gibbon species have to be considered carefully.

4.4.2.5 *Homo sapiens* – Human

The gross arrangement of the cancellous bone in human femora is well known (Fig. 4.15). The so called 'compressive trajectory' and the 'tensile trajectory' were identified in all specimens. Predominantly compressive loads are transmitted through the compressive trajectory, caused by body weight and muscle forces, pulling the femoral head into the acetabulum. During this loading condition the femoral head and neck are acting like a lever and give rise to tensile strains, which are transmitted via the tensile trajectory. A third trajectory laterodistally to the Ward's triangle was also discernible. The Ward's triangle is a region of diminished trabecular density, situated in the middle of the femoral neck, comprising the region framed by the former mentioned two trajectories. This latter trajectory is supposed to submit predominantly tensile strains (van Rietbergen et al. 2003). These tensile strains are hypothesized to be caused by the pull of the abductor and extensor muscles on the greater trochanter (M. gluteus medius, M. gluteus minimus, and M. piriformis, according to Platzer et al. (1986). Abductors and extensors are important for bipedal locomotion and in order to stabilize the posture during the stance phase during which only one leg is loaded. However, as bipedal locomotion consists mainly of walking, the loads caused by these muscles are steady but relatively low, while jumping and running cause much higher loads, up to multiple times the body weight. Therefore, and owing to species-specific postures the relative size of the greater trochanter is much larger in *Semnopithecus entellus* and *Papio hamadryas* compared to the rather short and stout greater trochanter of *Homo sapiens*. The trabecular alignment of the lesser trochanter in humans resembles that in *Papio hamadryas*. In the anterior half the trabeculae are inclined slightly laterally and in the posterior half they begin to reorient. Finally they are aligned in transverse direction in the posterior-most part of the lesser trochanter. As outlined above, this change of the trabecular alignment may result from differing loading directions of the flexor muscle, i.e. M. iliopsoas (Platzer et al. 1986), during different gaits.

4.4.2.6 *Pliopithecus vindobonensis* and *Paidopithex rhenanus*

The trabecular architectures of the femora of *Pliopithecus vindobonensis* and *Paidopithex rhenanus*

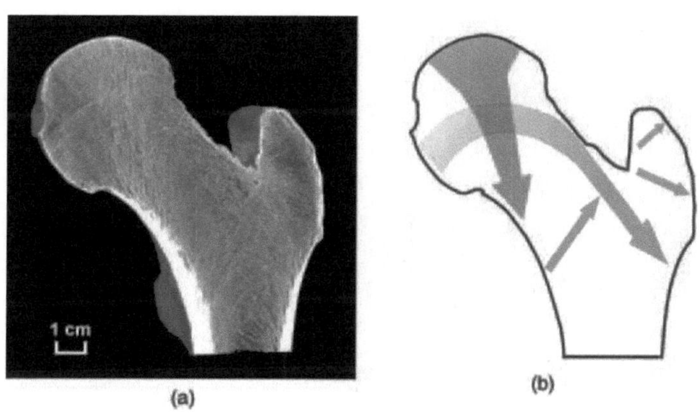

FIG. 4.15. (**a**) Ventral view of the trabecular architecture of the left femur of *Homo sapiens* (Senckenberg Research Institute, 21 ♂), Scherf (2007), (**b**) Outline of trabecular alignment (see Color Plates, Fig. 4.15).

are very similar (Figs 4.16, 4.17). Therefore, the following descriptions and comparisons are true for both fossil species. They show a quite unique trabecular architecture which is in its overall appearance not comparable to one of the extant species, while some features are similar to the extant specimens. In the proximal part of the femoral head a network of relatively thick and distant plate-like trabeculae is present, creating a cap-like structure. A similar arrangement was found in *Papio hamadryas* and *Semnopithecus entellus*. Concerning *Papio hamadryas*, this structure was interpreted as an indication of high loading conditions, especially during quadrupedal running, while its origin in *Semnopithecus entellus* remained unclear owing to the juvenile age of the sampled individuals. The *Papio hamadryas* specimens were all of adult age, while for *Pliopithecus vindobonensis* and *Paidopithex rhenanus* no definite statement about their ontogenetic age can be given. The assumption of an almost adult age is supported by the fact that no epiphyseal splits in the femora were found in the CT images, even though the teeth of individual I (O.E. 304) and II (1970/1397/22, 1970/1397/23) show a low wear state, probably indicative of a sub-adult age. Excluding ontogenetic processes, the biomechanical factors causing the described structure must have been the same in all specimens, indicating a frequent and relatively high loading of the femoral head in *Pliopithecus vindobonensis* and *Paidopithex rhenanus*, probably in connection with fast quadrupedal locomotion. The intensity of this feature is supposed to be linked to actual loads. Comparing this intensity in *Papio hamadryas* and in the two fossil species, the loads acting on the femora of the two fossil species must have been less vigorous. Below the cap-like structure a dense network of thinner trabeculae is present in *Pliopithecus vindobonensis* and *Paidopithex rhenanus*.

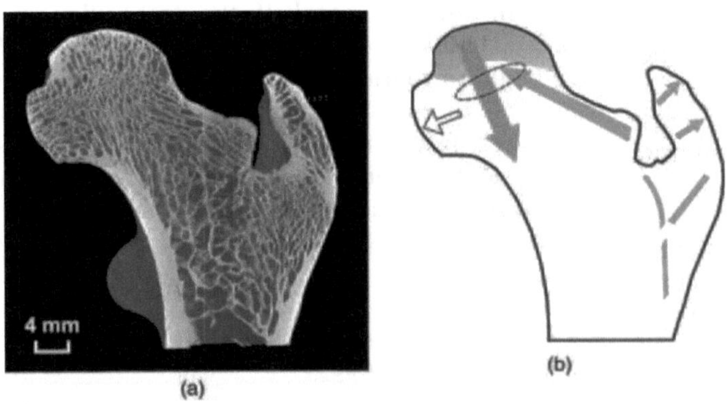

FIG. 4.16. (**a**) Ventral view of the trabecular architecture of the left femur of *Pliopithecus vindobonensis* (Naturhistorisches Museum Basel, O.E. 559), Scherf (2007), (**b**) Outline of trabecular alignment (see Color Plates, Fig. 4.16).

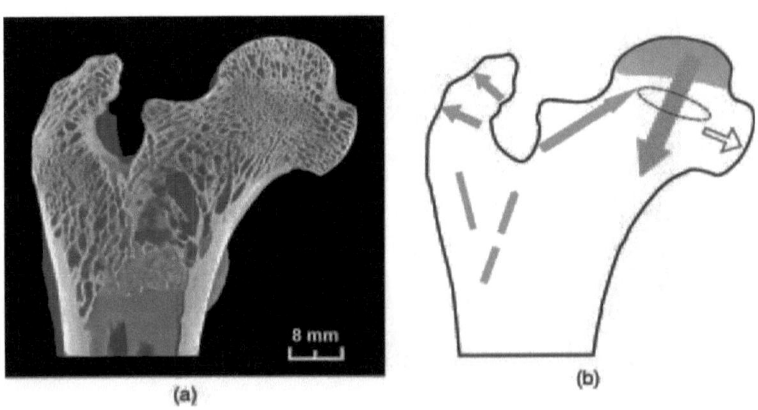

FIG. 4.17. (**a**) Ventral view of the trabecular architecture of the right femur of *Paidopithex rhenanus* (Hessisches Landesuseum Darmstadt, Din 45), Scherf (2007), (**b**) Outline of trabecular alignment (see Color Plates, Fig. 4.17).

A comparable feature was found in *Hylobates lar moloch*, but the loading condition giving rise to this feature remains unclear.

Two trajectories are present in the femoral head and neck region of the fossil species. The first trajectory runs from the proximal part of the inferior cortex of the femoral neck to the proximal end of the femoral head. On its course, it crosses the cap-like structure and the adjoining dense network of thin trabeculae described above. The second trajectory extends sub-parallel to the superior cortex of the femoral neck and ends in the mid part of the femoral head at the dense network of thin trabeculae. The relative size of the greater trochanter of the fossils is similar to *Hylobates*. Some features of the trabecular architecture in the superior half of the greater trochanter are comparable to the extant species but in its inferior half unique features of the fossil specimens occur. Therefore, it is not possible to infer the possible loading conditions of the greater trochanter in the fossil species by simple comparison with the extant species. *Pliopithecus vindobonensis* has a prominent lesser trochanter, similar to *Papio hamadryas* but while the distal part is emphasized in *Papio hamadryas*, *Pliopithecus vindobonensis* shows a rather homogeneously shaped lesser trochanter. The trabeculae of the lesser trochanter are anteriorly inclined towards the lateral side, while posteriorly they align in transverse direction.

Considering these observations (Table 4.3) it becomes obvious that the loading conditions of the proximal femur of *Pliopithecus vindobonensis* and *Paidopithex rhenanus* are different compared to those in *Alouatta seniculus* and *Hylobates* to which they have been compared previously, even though some similarities to *Hylobates* exist. Overall it appears that the trabecular architecture in the fossil species was able to transmit higher loads compared to those caused by brachiation or semi-brachiation. The trajectories and the rather heterogeneous cancellous network in *Pliopithecus vindobonensis* and *Paidopithex rhenanus*, compared to the rather homogeneous cancellous network of *Alouatta seniculus* and *Hylobates*, support this hypothesis. The trabecular architecture of *Semnopithecus entellus* and *Homo sapiens* shows fewer similarities to the two fossil species.

On the basis of these comparisons it seems reasonable to conclude that *Pliopithecus vindobonensis* and *Paidopithex rhenanus* were rather locomotor generalists, occasionally performing some load intensive kinds of locomotion causing the relatively well aligned cancellous network and the structure of relatively thick trabeculae in the superior part of the femoral head. However, the gross architecture shows no signs for specialized, load intensive locomotion like in *Papio hamadryas* and *Semnopithecus entellus*.

4.5 Closing Remarks

High resolution computed tomography is an optimal tool to investigate cancellous bone structures of long bones. As a non-destructive method it is quite appropriate for fossil material and the specimen remains intact and can be subjected to further investigations. The high resolution CT images reveal that different locomotor habits cause different cancellous bone architectures in the proximal femur, which can be defined by their characteristic features. Two types in gross architecture can be distinguished. A homogeneous type, with a less directed trabecular network, as presented by *Alouatta seniculus*, *Hylobates seniculus*, and *H. lar moloch*, and a heterogeneous type with a well directed trabecular network, as in *Homo sapiens*, *Semnopithecus entellus*, and *Papio hamadryas*. The homogeneous type goes along with low loading conditions acting on the hind limbs from various directions, as they occur during climbing. The heterogeneous type is associated with specialized types of locomotion, subjecting the hind limbs to relatively high loads.

The present study yields new insight into the complex relation of locomotor loading and bone structure and forms a new basis for investigating cancellous bone architecture. A future combination of the high resolution CT analysis with other fields of study comprising locomotor behaviour, muscular systems, muscle insertions, composition of bone on micron level, and FEM (Finite Element Method) studies will contribute to the knowledge of locomotor biomechanics and the generation of comprehensive models for different modes of habitual locomotion. Locomotion models of extant species developed on the basis of such combined approaches, together with comparative analyses of the outer bone morphology, provide a well founded basis for locomotor interpretations of fossil species.

Table 4.3. Outline of trabecular arrangement according to different gross loading types (see Color Plates, Table 4.3).

High hind limb loading	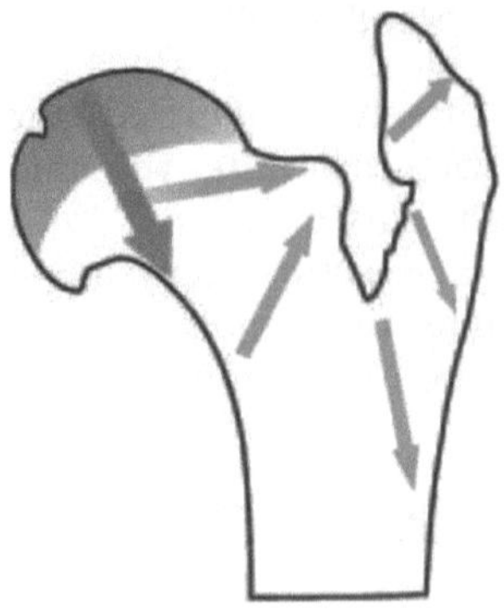 *Papio hamadryas*	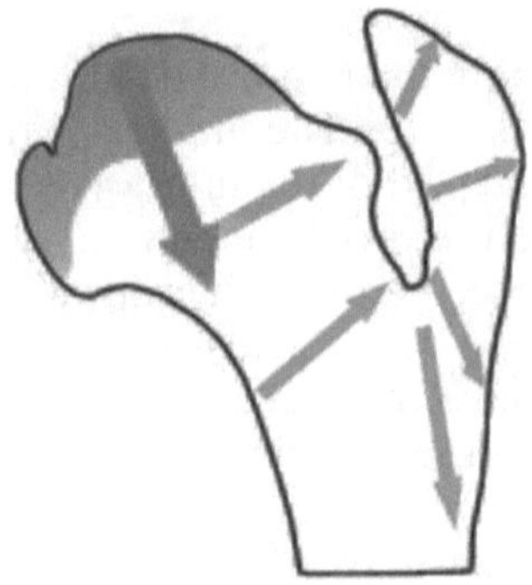 *Semnopithecus entellus*	
Medium hind limb loading	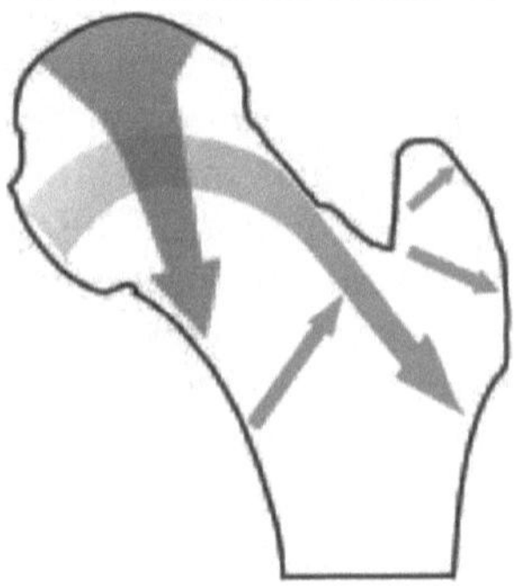 *Homo sapiens*		
Low hind limb loading	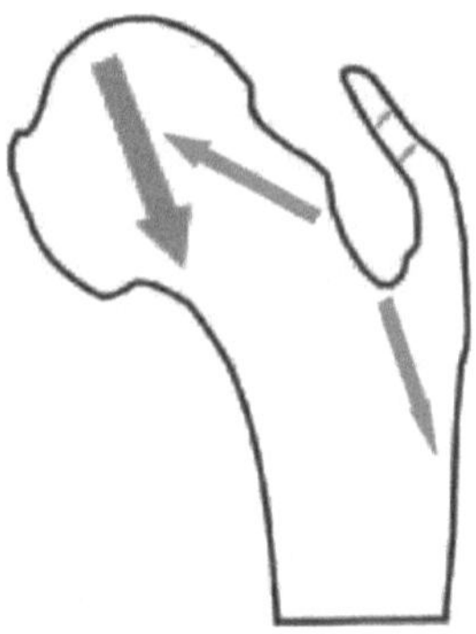 *Alouatta seniculus*	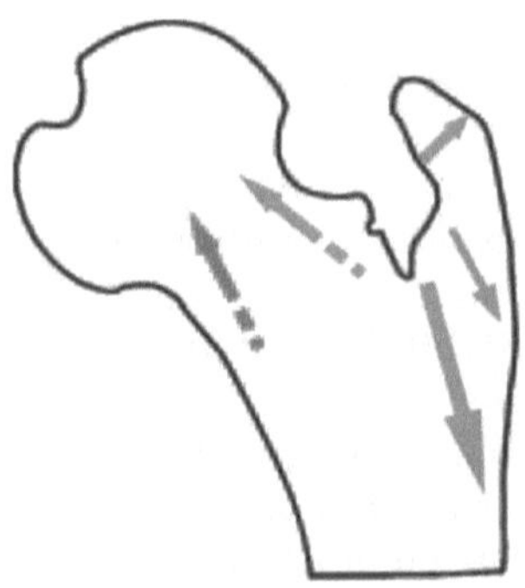 *Hylobates syndactylus*	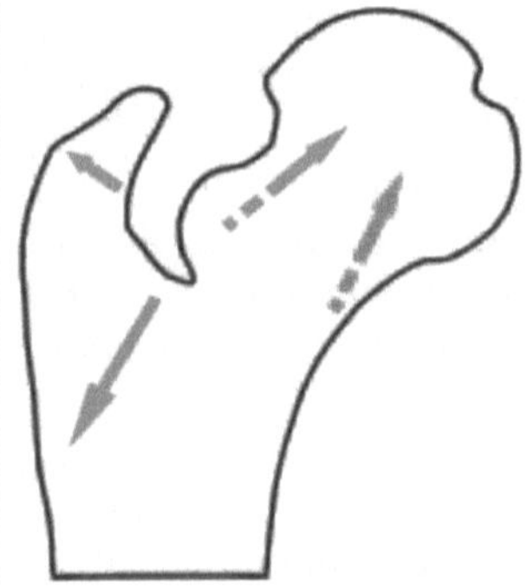 *Hylobates lar moloch*
Fossil species	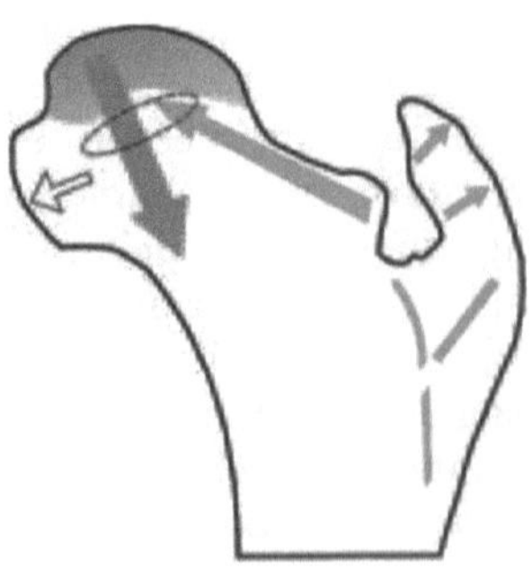 *Pliopithecus vindobonensis*	 *Paidopithex rhenanus*	

Acknowledgement

I thank especially Roland Frey and Hideki Endo for their kind invitation to contribute to this volume and their substantial inputs and improving of the manuscript and organizing the symposium on the combination of classical methods with CT analysis in evolutionary morphology at the 9th International Mammalogical Congress.

I thank Friedemann Schrenk for enabling this study and I thank Bernd Herkner for his careful advice. I am very grateful to Gabriele Gruber, Hessian State Museum Darmstadt (Germany), Christian Meyer and Burkart Engesser, Natural History Museum Basel (Switzerland), and Gudrun Höck, Natural History Museum Vienna (Austria) for giving me the chance to investigate *Paidopithex rhenanus* and *Pliopithecus vindobonensis*, and further Gerhard Storch, Senckenberg Research Institute (Frankfurt/Main, Germany), Helmut Wicht, Dr. Senckenbergische Anatomie (Frankfurt/Main, Germany), Dietrich von Knorre, Phyletic Museum Jena (Germany), and Harald Schliemann and Nelson Mascarenhas, Institute of Zoology of the University of Hamburg (Germany) for providing the extant specimens. I thank Jens Lorenz Franzen for his helpful comments to the Eppelsheim site. I also thank Allen Goodship, Claudia Barelli, Larissa Swedell, and Paul Winkler, who very kindly allowed me to use some of their images. And special thanks to Jeremy Tausch and Katerina Harvati, who helped me to find the right words and also to Christine Hemm, Elke Pantak-Wein, Annika Hebs, Uta Schwarz, and Michael Gasch for their great assistance. This study was supported by a scholarship for women in natural and engineering sciences of the Darmstadt University of Technology, donated by the Hessian Ministry of science and art, the Ermann Foundation, and the Senckenberg Research Institute (Frankfurt/Main, Germany), for which I like to thank especially the former director Fritz F. Steininger.

References

Adler C-P (1998) Knochenkrankheiten. Springer-Verlag, Berlin

Andrews P, Harrison T, Delson E, Bernor RL, Martin L (1996) Distribution and biochronology of european and southwest asian Miocene catarrhines. In: Bernor RL, Fahlbusch V, Mittmann H-W (eds) The evolution of western eurasian neogene mammal faunas. Columbia University Press, New York, pp 168–207

Begun DR (1992) Phyletic diversity and locomotion in primitive european hominids. Am J Phys Anthropol 87:311–340

Biewener AA (1989) Scaling body support in mammals: Limb posture and muscle mechanics. Science 245:45–48

Borah B, Gross GJ, Dufresne TE, Smith TS, Cockman MD, Chmielewski PA, Lundy MW, Hartke JJ, Sod EW (2001) Three-dimensional microimaging (MrμI and μCT), finite element modeling, and rapid, prototyping provide unique insight into bone architecture in osteoporosis. *Anat Rec (Part B, New Anat)* 265:101–110

Borland R (1991) Menschen und Tiere – Bei den Pavianen. ORF

Carpenter CR (1934) A field study of the behaviour and social relations of howling monkeys (*Alouatta palliata*). *Comparative Psychology Monographs X*. The John Hopkins Press, Baltimore, MD

Carpenter CR (1940) A field study in Siam of the behavior and social relations of the gibbon (*Hylobates lar*). Reprinted in *Naturalistic behavior of non-human primates* (Carpenter CR, 1964). The Pennsylvania State University Press, University Park, IL

Chambers TJ, Evans M, Gardner TN, Turner-Smith A, Chow JWM (1993) Induction of bone formation in rat tail vertebrae by mechanical loading. Bone Miner 20:167–178

Collet J-Y, Vienne G (1986–1989a) Die Affen – Neuweltaffen in Südamerika. Bayrischer Rundfunk 1990

Collet J-Y, Vienne G (1986–1989b) Die Affen – Paviane und Mantelaffen in Afrika. Bayrischer Rundfunk 1990

Compston JE (1994) Connectivity of cancellous bone: assessment and mechanical implications. Bone 15:463–466

Dempster DW (1992). Bone remodeling. In: Coe FL, Favus MJ (eds) Disorders of bone and mineral metabolism. Raven Press, New York, pp 355–380

Daxner-Höck G (1998) Säugetiere (Mammalia) aus dem Karpat des Korneuburger Beckens – 3. Rodentia und Carnicora. Beiträge Paläontol 23:367–407

Day MH (1979) The locomotor interpretation of fossil primate postcranial bones. In: Morbeck ME, Preuschoft H, Gomberg N (eds) Environment, behaviour, and morphology: dynamic interactions in primates. Fischer, New York, pp 245–258

Duda GN (1996) Influence of muscle forces on the internal loads in the femur during gait. PhD thesis, Technical University Hamburg-Harburg. Shaker Verlag, Aachen

Fajardo RJ, Müller R (2001) Three-dimensional analysis of nonhuman primate trabecular architecture using micro-computed tomography. Am J Phys Anthropol 115:327–336

Fischer J (1961) Vergleichend-anatomische Untersuchungen über die Hüft- und Oberschenkelmuskulatur von Papio leucophaeus (CUVIER 1807) und Mensch. PhD thesis, Medical Academy Düsseldorf

Fleagle JG (1976) Locomotion and posture of the malayan Siamang and implications for hominoid evolution. Folia Primatol 26:245–269

Fleagle JG (1980) Locomotion and posture. In: Chivers DJ (ed.) Malayan forest primates. Plenum Press, New York, pp 191–207
Fleagle JG (1983). Locomotor adaptations of Oligocene and Miocene hominoids and their phyletic implications. In: Ciochon RL, Corruccini RS (eds) New interpretations of ape and human ancestry. Plenum Press, New York, pp 301–324
Franzen JL, Fejfar O, Storch G (2003) First micromammals (mammalia, Soricomorpha) from the Vallesian (Miocene) of Eppelsheim, Rheinhessen (Germany). Senck leth 83:95–102
Frost HM, Ferrretti JL, Jee WSS (1998) Perspectives: some roles of mechanical usage, muscle strength, and the mechanostat in skeletal physiology, disease, and research. Calcif Tissue Int 62:1–7
Goldstein SA, Matthews LS, Kuhn JL, Hollister SJ (1991) Trabecular bone remodeling: an experimental model. J Biomech 24:135–150
Goodship AE, Lanyon LE, McFie H (1979) Functional adaptation of bone to increased stress. J Bone Joint Surg [Am] 61:539–546
Goulet RW, Goldstein SA, Ciarelli MJ, Kuhn JL, Brown MB, Feldkamp LA (1994) The Relationship between the structural and orthogonal compressive properties of trabecular bone. J Biomech 27:375–389
Grand TI (1968a) Functional anatomy of the upper limb. Bibl primatol 7:104–125
Grand TI (1968b) The functional anatomy of the lower limb of the Howler Monkey (*Alouatta caraya*). Am J Phys Anthropol 28:163–182
Guldberg RE, Hollister SJ (1995) Influence of loafing on the tissue modulus of trabecular bone: a combined experimental and microstructural modeling approach. Adv Bioeng ASME BED-31:157–158
Guldberg RE, Caldwell NJ, Guo XE, Goulet RW, Hollister SJ, Goldstein SA (1997a) Mechanical stimulation of tissue repair in the hydraulic bone chamber. J Bone Miner Res 12:1295–1302
Guldberg RE, Richards M, Caldwell NJ, Kuelske CL, Goldstein SA (1997b) Trabecular bone adaptation to variations in porous-coated implant topology. J Biomech 30:147–153
Günther MM (1989) Funktionsmorphologische Untersuchungen zum Sprungverhalten an mehreren Halbaffenarten. PhD thesis, Free University of Berlin
Hall KRL (1962) Numerical data, maintenance activities and locomotion of the wild Chamca Baboon, *Papio ursinus*. Proc Zool Soc Lond 139:181–220
Huiskes R (1997). Simulation of self-organization and functional adaptation in bone. In: Schneider E (ed.) Biomechanik des menschlichen Bewegungsapparates (*Der Unfallchirurg/Hefte* 261). Springer-Verlag, Berlin, pp 299–320
Huiskes R, Ruimerman R, van Lenthe GH, Janssen JD (2000) Effects of mechanical forces on maintenance and adaptation of form in trabecular bone. Nature 405:704–706
Krieg H (1928) Schwarze Brüllaffen (*Alouatta caraya* Humboldt). Tagebuch-Aufzeichnungen auf der Deutschen Chaco-Expedition. Z Saeuget II:119–132
Kummer B (1959) Bauprinzipien des Säugerskeletts. Georg Thieme Verlag, Stuttgart
Langdon JH (1986) Functional morphology of the Miocene hominoid foot. Contrib Primatol 22:1–225
Lanyon LE (1974) Experimantal support for the trajectorial theory of bone structure. J Bone Joint Surg [Br] 56:160–166
Lanyon LE (1981) Locomotor loading and functional adaptation in limb bones. Symp Zool Soc Lond 48:305–329
van der Linden JC, Birkenhäger-Frenkel DH, Verhaar JAN, Weinans H (2001) Trabecular bone's mechanical properties are affected by its non-uniform mineral distribution. J Biomech 34:1573–1580
MacLatchy L, Müller R (2002) A comparison of the femoral head and neck trabecular architecture of Galago and Perodictius using micro-computed tomography (μCT). J Hum Evol 43:89–105
Martill DM (1991) Bones as stones: the contribution of vertebrate remains to the lithologic record. In: Donovan SK (ed.) The process of fossilization. Belhaven Press, London, pp 270–292
McHenry H, Corruccini RS (1976) Affinities of tertiary hominoid femora. Folia Primatol 26:139–150
McNeill Alexander R (1985) Body support, scaling, and allometry. In: Hildebrand M, Bramble DM, Liem KF, Wake DB (eds) Functional vertebrate morphology. Belknap Press, Cambridge, MA, pp 26–37
Morgan EF, Keaveny TM (2001) Dependence of yield strain of human trabecular bone on anatomic site. J Biomech 34:569–577
Mullender MG, Huiskes R (1995) Proposal for the regulatory mechanism of Wolff's law. J Orthop Res 13:503–512
Mullender MG, Huiskes R, Weinans H (1994) A physiological approach to the simulation of bone remodeling as a self-organizational control process. J Biomech 27: 1389–1394
Müller R, Van Campenhout H, Van Damme B, Van der Perre G, Dequeker J, Hildebrand T, Rüegsegger P (1998) Morphometric analysis of human bone biopsies: a quantitative structural comparison of histological sections and micro-computed tomography. Bone 23:59–66
Napier JR (1976) Primate locomotion. Oxf Biol Readers 41:3–16
Nikolei J (2002) Lokomotionsökologie adulter Hanuman Languren (*Semnopithecus entellus*) in einem saisonalen Waldhabitat in Ramnagar, Südnepal. PhD thesis, Free University of Berlin
Ott SM (1996). Theoretical and methodical approach. In: Bilezikian JP, Raisz LG, Rodan GA (eds) Principles of bone biology. Academic Press, San Diego, CA, pp 231–241
Parfitt AM (1983). The physiologic and clinic significance of bone histomorphometric data. In: Recker RR (ed.) Bone histomorphometry: techniques and interpretatoin. CRC Press, Inc, Boca Raton, FL, pp 143–223
Pauwels F (1965) Gesammelte Abhandlungen zur funktionellen Anatomie des Bewegungsapparates. Springer-Verlag, Heidelberg
Pauwels F (1980) Biomechanics of the locomotor apparatus. Springer-Verlag, Heidelberg

Platzer W, Kahle W, Leonhardt H (1986) Taschenatlas der Anatomie. Thieme Verlag, Stuttgart

Preuschoft H (1988) Kleine Menschenaffen oder Gibbons. In: Grzimek B (ed.) Grzimeks Enzyklopädie der Säugetiere, Vol. 2. Kindler Verlag, München, pp 328–356

Rafferty KL (1998) Structural design of the femoral neck in primates. J Hum Evol 34:361–383

van Rietbergen B, Huiskes R, Eckstein F, Rüegsegger P (2003) Trabecular bone tissue strains in the healthy and osteoporotic human femur. J Bone Miner Res 18: 1781–1788

Rose MD (1994) Quadrupedalism in some Miocene catarrhines. J Hum Evol 26:387–411

Ryan TM, Ketcham RA (2002a) Femoral head trabecular bone structure in two omomyid primates. J Hum Evol 42:241–263

Ryan TM, Ketcham RA (2002b) The three-dimensional structure of trabecular bone in the femoral head of strepsirrhine primates. J Hum Evol 43:1–26

Salamone LM, Glynn N, Black D, Epstein RS, Palermo L, Meilahn E, Kuller LH, Cauley JA (1995) Body composition and bone mineral density in premenopausal and early perimenopausal women. J Bone Miner Res 10:1762–1768

Scherf H (2007) Locomotion-related femoral trabecular architectures in Primates (*Paidopithex rhenanus, Pliopithecus vindobonensis*). PhD thesis, Darmstadt University of Technology. http://elib.tu-darmstadt.de/diss/000797

Scherf H, Koller B, Schrenk F (2005) Locomotion related structures in the femoral trabecular architecture of Primates and Insectivores. Senck biol 85:101–112

Simon M, Sauerwein C, Tisenau I, Burdairon S (2001) Felxible 3D-Computertomographie im RayScan 200. *DGZfP-Jahrestagung 2001, Zerstöringsfreie Materialprüfung, Berlin 21.–23. Mayi 2001*, 20.08.2006. http://www.ndt.net/article/dgzfp01/papers/v49/v49.htm

Simons EL, Fleagle J (1973) The history of extinct gibbon-like primates. Gibbon and Siamang 2:121–148

Stenström M, Olander B, Letho-Axelius D, Madsen JE, Nordsletten L, Carlsson GA (2000) Bone mineral density and bone structure parameters as predictors of bone strength: as analysis using computerized microtomography and gastrectomy-induced osteopenia in the rat. J Biomech 33:289–297

Swindler DR, Wood CD (1973) An atlas of primate gross anatomy. University of Washington Press, Seattle

Szalay FS, Delson E (1979) Evolutionary history of the primates. Academic Press, New York

Tsubota K, Adachi T, Tomita Y (2002) Functional adaptation of cancellous bone in human proximal femur predicted by trabecular surface remodeling simulation toward uniform stress state. J Biomech 35:1541–1551

Vereecke E (2006) The functional morphology and bipedal locomotion of Hylobates lar and its implications for the evolution of hominin bipedalism. PhD thesis, University Antwerpen

Vogel C, Winkler P (1988) Schlank- und Stummelaffen. In: Grzimek B (ed.) Grzimeks Enzyklopädie der Säugetiere, Vol. 2. Kindler Verlag, München, pp 296–325

Weaver JK, Chalmers J (1966) Cancellous bone: Its strength and changes with aging and an evaluation of some method for measuring its mineral content. J Bone Joint Surg [AM] 48:289–298

Welker C, Schäfer-Witt C (1988) Kapuzinerartige. In: Grzimek B (ed.) Grzimeks Enzyklopädie der Säugetiere, Vol. 2. Kindler Verlag, München, pp 122–177

Welten DC, Kemper HCG, Post GB, van Mechelen W, Twisk J, Lips P, Teule GJ (1994) Weight-bearing activity during youth is a more important factor for peak bone mass than calcium intake. J Bone Miner Res 9:1089–1096

Whalen RT, Carter DR, Steele CR (1988) Influence of physical activity on the regulation of bone density. J Biomech 21:825–837

Whitehouse WJ, Dyson ED (1974) Scanning electron microscope studies of trabecular bone in the proximal end of the human femur. J Anat 118:417–444

Whitehouse WJ, Dyson ED, Jackson CK (1971) The scanning electron microscope in studies of trabecular bone from a human vertebral body. J Anat 108:481–496

Williams JL, Lewis JL (1982) Properties and an anisotropic model of cancellous bone from the proximal tibial epiphysis. J Biomech Eng 104:50–56

Wolff J (1892) Das Gesetz der Transformation der Knochen. Schattauer, Stuttgart; Reprint 1991, Hirschwald, Berlin

Zapfe H (1960) Die Primatenfunde aus der miozänen Spaltenfüllung von Neudorf an der March (Děvínská Nová Ves), Tschechoslowakei. Schweiz Palaeontol Abh 78:1–293

5 Three-dimensional Imaging of the Manipulating Apparatus in the Lesser Panda and the Giant Panda

Hideki Endo[1*], Natsuki Hama[2], Nobuharu Niizawa[3], Junpei Kimura[4], Takuya Itou[5], Hiroshi Koie[6], and Takeo Sakai[5]

5.1 Introduction

The unique manipulation mechanism has been morphologically examined in the lesser panda (*Ailurus fulgens*) (Endo et al. 2001b). Since the manipulating radial sesamoid bone is weakly attached to the palm region, our data suggest that the lesser panda can adjust the supporting bar of the radial sesamoid bone during the seizing action. The lesser panda and the giant panda (*Ailuropoda melanoleuca*) have independently evolved a specialized manipulation mechanism using the radial sesamoid bone. The skillful manipulating system consists of a huge radial sesamoid bone in the giant panda, and this species can strongly seize bamboo stems unlike the lesser panda (Endo et al. 1999a,c, 2001a). Our results have clarified the parallel evolution of the extraordinary manipulating system in the lesser panda and the giant panda. However, the three-dimensional simulation of the radial sesamoid bone had not been undertaken in the lesser panda. So far, the detailed movements of the bones in the carpal and palm areas had remained unclear even in the giant panda. Here the manipulating action using the additional bones in the two species was detailed by three-dimensional CT imaging methods.

5.2 Between the Two Periods of the CT Scanning and the Related Techniques

The digitalized techniques enable us the visualization of the animal body in the three-dimensional techniques of the computer science. This situation is not traditional but appeared in the 1990's. In this first period of the 1990's, from the needs of the clinical image examinations, the CT systems were only adapted to print analogous photographs or negative films. The system of the early period was not so developed to quantitatively examine the functions of the animal body. However, for example, we clarified the manipulating function of the giant panda (Endo et al. 1996, 1999a,c, 2001a, 2003a), and the digging mechanisms of the aardvark (Endo et al. 2003b) using the old-fashioned CT systems of the early 1990's.

The next generation of the CT system is actually supported by the customers of the Image analysis fields. The present CT systems established in the digitalized technologies enable us to connect the CT data with the computer systems generalized in the hospitals, laboratories and offices.

[1]Section of Morphology, Primate Research Institute, Kyoto University, Inuyama, Aichi 484-8506, Japan, [2]Kobe Municipal Oji Zoo, Hyogo 657-0838, Japan, [3]Animal Medical Center, College of Bioresource Sciences, Nihon University, Kanagawa 252-8510, Japan, [4]Department of Veterinary Anatomy and Cell Biology, College of Veterinary Medicine, Seoul National University, Seoul 151-742, Korea, [5]Department of Preventive Veterinary Medicine and Animal Health, College of Bioresource Sciences, Nihon University, Kanagawa 252-8510, Japan, [6]Department of Veterinary Internal Medicine, College of Bioresource Sciences, Nihon University, Kanagawa 252-8510, Japan, *Present address: The University Museum, The University of Tokyo, Bunkyo-ku, Tokyo 113-0033, Japan

5.3 The Relationships of the Lesser Panda and the Giant Panda

The lesser panda (*Ailurus fulgens*) and the giant panda (*Aliluropoda melanoleuca*) have independently evolved the specialized manipulation mechanism using the radial sesamoid bone as indicated by molecular phylogenetic studies (Delisle and Strobeck 2005; Fulton and Strobeck 2006). The lesser panda belongs to the Musteloidea and the giant panda is positioned in the Family Ursidae.

The anatomists have considered the radial sesamoid bone of the giant panda as an example of the most extraordinary manipulating structures in mammalian evolution (Pocock 1939; Wood-Jones 1939a,b; Davis 1964; Gould 1978; Beijing Zoo et al. 1986), and recently we described the system by the methods of macroscopic anatomy (Endo et al. 1996, 1999b). In addition, we pointed out that the radial sesamoid may not function as an active manipulating apparatus based upon computed tomography (CT) and magnetic resonance imaging (MRI) results and that the double pincer-like functional units and hammock-like muscles function as effective seizing system in the giant panda (Endo et al. 1999a,c, 2001a; 2003a).

In the lesser panda, however, the radial sesamoid bone functions as a simple supporting bar (Endo et al. 2001b). The radial sesamoid bone consists of bone and cartilage, and it is passively moved in the regions of the palm and wrist confirmed by naked eye observations and histological examinations (Endo et al. 2001b). Since then, the functional morphological simulation of the radial sesamoid bone has never been undertaken. In contrast to the giant panda, the functional role of the radial sesamoid has remained unclear in the lesser panda. In this study therefore, it has been expected that the manipulating action related to the radial sesamoid bone could be reconstructured by three-dimensional CT imaging in the lesser panda.

5.4 The Way of CT Scanning and Analysis

A dead body of the lesser panda (*Ailurus fulgens*) of Kobe Municipal Oji Zoo was subjected to three-dimensional image analysis. The basic biological data of the individual is female and 19 years old. We did virtual serial sections of the distal part of the forelimbs by multiple CT scanner (Computed tomography) (Toshiba Aquilion16: Toshiba Medical Systems, Japan) from distal to proximal planes at 0.5 mm thickness without gap. Both opened and manipulating states were simulated and scanned in this forearm. The three-dimensional images were reconstructed from the series of CT sections to observe the position and movement of the radial sesamoid bone and the other related bones of the hand. We applied the Voxel Transmission (Volume Rendering) techniques to visualize the images of bone movements through a three-dimensional image analyzing system (AZE Virtual Place: AZE Corporation, Tokyo, Japan). The vector analyses were carried out to quantitatively reveal the position of each bone in the three dimensional coordinates by the same software. The data collecting way has also been explained in our article (Endo et al. 2007a,b). In the giant panda, the radial sesamoid and the first metacarpal bones were virtually sectioned by the same methods and the three-dimensional reconstruction was carried out. The bone specimen of the giant panda has been stored in the National Science Museum, Tokyo, and its museum number is NSMT-M31395.

5.5 The Analysis of the Data and the Observation

The three dimensional reconstructed images of both opened and manipulating states of the right hand were shown at some aspects in the lesser panda (Figs 5.1–5.3). These figures also include those images taken before removing the soft parts. The radial sesamoid appears as a cone-shaped bone at the radial side of the wrist, and does not have an articulation surface to the first metacarpal. It is attached to the radial carpal. However, we can observe that the position and direction of the radial sesamoid bone against the radial carpal changes between the two states.

In a previous report (Endo et al. 2007) a vector analysis was carried out to clarify the role of each bone in the manipulating action. The four planes were set in a three-dimensional vector space as follows: (1) The plane representing the palm; (2) The plane representing the radius and ulna; (3) The plane representing the radial sesamoid;

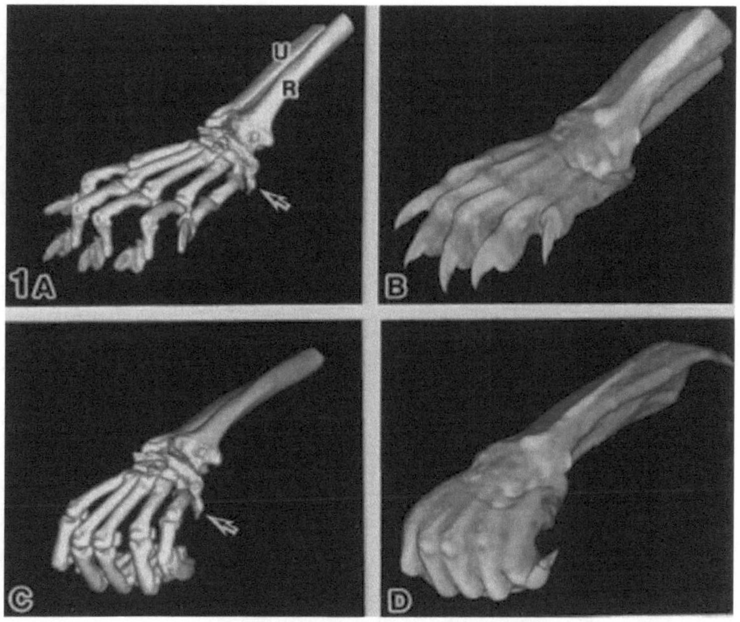

FIG. 5.1. Three-dimensional reconstructed images of the skeleton of the lesser panda. Medial and distal aspect of the right hand. Opened state (**A** and **B**) and manipulating state (**C** and **D**). Only skeleton (**A** and **C**) and the skeleton with soft parts (**B** and **D**) are seen. The radial sesamoid (*arrow*) is shown in the wrist area. *R*, radius; *U*, ulna.

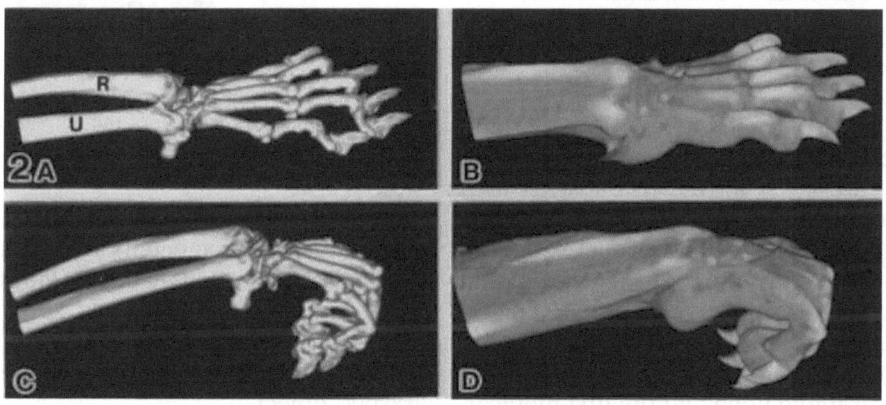

FIG. 5.2. Lateral aspect of the right hand. Symbols are the same as in Fig. 5.1.

and (4) The plane representing the accessory carpal. The angles composed by the four planes were calculated from the data of the normal vectors of each plane. As a result of the vector analysis, the radial sesamoid plane can move independently from the metacarpals plane during the manipulating action (Endo et al. 2007).

We concluded that the manipulation mechanism of the lesser panda (Endo et al. 2001b) is different from that of the giant panda (Endo et al. 2001a; 2003a), although both species possess a relatively large radial sesamoid bone (Davis 1964). Although the wrist joint flexes during the manipulating action in the lesser panda, the flexing angle of the wrist joint is small compared to the giant panda (Endo et al. 1999a,c, 2001a, 2007). The large sesamoid bone is not fixed to the first metacarpal bone in the lesser panda, and the structure is weakly positioned by the tendons of the Musculus abductor pollicis longus, M. abductor pollicis brevis and M. opponens pollicis, Aponeurosis palmaris and Flexor retinaculum within the palm

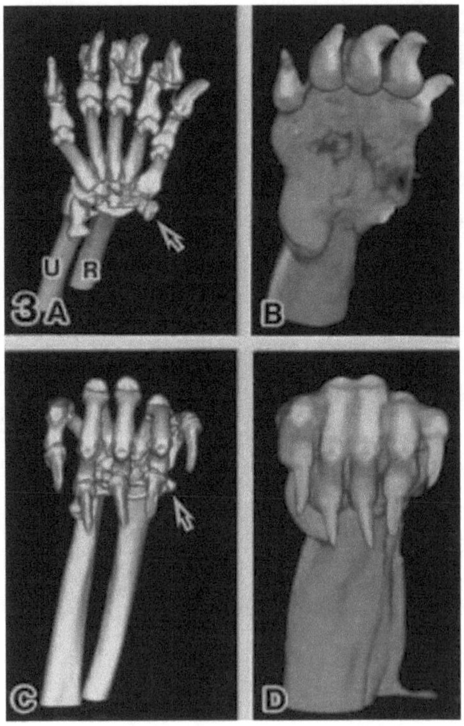

FIG. 5.3. Distal and palmer aspects of the right hand. Symbols are the same as in Fig. 5.1.

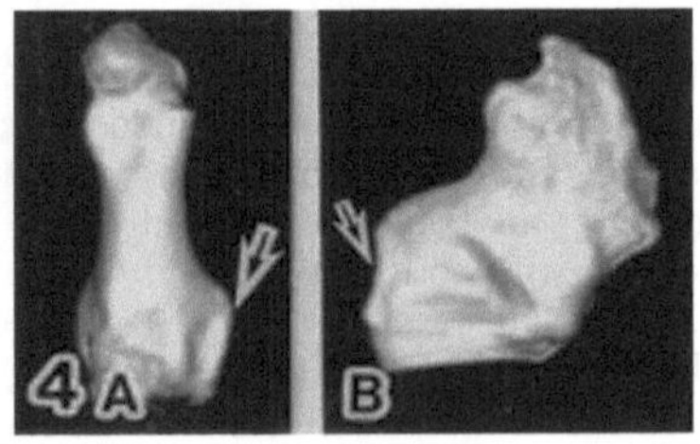

FIG. 5.4. The first metacarpal (**A**) and the radial sesamoid (**B**) of the giant panda. Dorsal aspect of the left side. The articulations (*large arrows*) between the radial sesamoid and the first metacarpal are seen. The connecting surface (*small arrow*) to the radial carpal is also shown in the radial sesamoid.

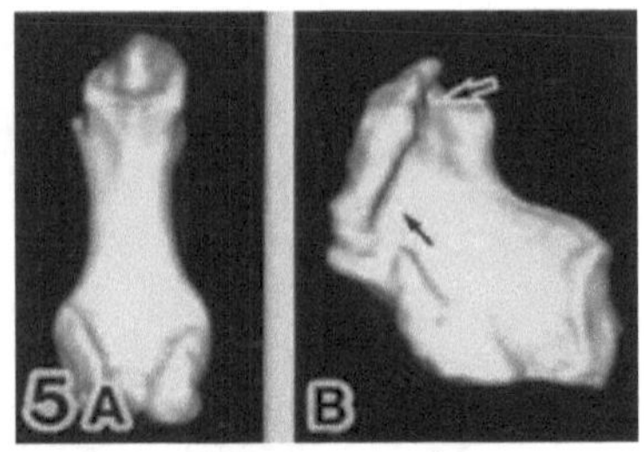

FIG. 5.5. The first metacarpal (**A**) and the radial sesamoid (**B**) of the giant panda. Palmar aspect of the left side. The protruding margin (*arrows*) of the medial side is confirmed in radial sesamoid.

(Endo et al. 2001b). It could be concluded that the radial sesamoid bone is moved in the wrist area according to the size and shape of grasped objects in the lesser panda, whereas the radial sesamoid bone is strongly fixed to the first metacarpal and does not change its position against the first metacarpals in the giant panda (Endo et al. 1999a–c, 2001a, 2003a).

The radial sesamoid bone and the first metacarpal of the giant panda are shown (Figs 5.4,5.5). The radial sesamoid bone is extraordinary large. The immobility of the articulation surfaces are confirmed (Figs 5.4,5.5). The single complex model indicated in the early analysis of the giant panda (Endo et al. 1999a–c, 2001a, 2003a) is supported these digitalized images. The distal and medial margin of the radial sesamoid bone is sharply bent (Fig. 5.5) and it suggests that the margin of the bone may act as a supporter for the seized objects.

We had the phalanges and metacarpals extended and flexed from each proximal joint in the three-dimensional simulation (Endo et al. 2007). In the simulation, the radial sesamoid bone easily moved in the wrist and changed its angle to the first metacarpal. We suggest that this findings might result from passive tensions of the soft tissues in the manipulating action. The radial sesamoid bone can move easily to be fitted to the form of the grasped object.

5.6 The Data of Molecular Phylogeny Related the Lesser Panda and the Expected Three-Dimensional Morphological Comparisons

The manipulation mechanism is functional-morphologically different between the lesser panda and the giant panda, although the two species both use the radial sesamoid bone as a manipulating apparatus, as has been documented by three-dimensional imaging in the previous (Endo et al.

2007) and present studies. The monophyly of the lesser panda and the skunks was demonstrated by Delisle and Strobeck (2005). The skunks separated first, and next the lesser panda made an evolutionary branching point from the Musteloidea (Fulton and Strobeck 2006). Since the giant panda represents an Ursidae species these molecular data support the convergence of the seizing mechanisms of the radial sesamoid bone between the lesser panda and the giant panda.

Davis (1964) detailed the morphological differences of the radial sesamoid bone among the giant panda, lesser panda, raccoon and some related carnivores. It was concluded that the radial sesamoid bone of the giant panda has been extraordinarily developed in comparison with its body size. However, the author described the large radial sesamoid bone of the lesser panda as noteworthy. Although the genetic theory had not been refined in the 1960s, the evolutionary relationships between large radial sesamoid bone and the two species of pandas attracted the morphologists. The movement of the palm and wrist regions including the radial sesamoid bone could be quantitatively examined in the lesser panda (Endo et al. 2007). Our findings and the latest molecular genetic results point out that these two phylogenetically distinct groups independently evolved the obviously different mechanisms using the same elements, the radial sesamoid and the accessory carpal bones.

The present standard of the CT analysis could not be applied to the dead body of the giant panda (Endo et al. 1996, 1999a,c, 2001a, 2003a). Although we could partially confirm each bone movement in the manipulating action of the lesser panda, the mechanisms of the bone movement in the giant panda could not be compared with those in the lesser panda in three-dimensional digitalized space. The comparative quantitative image analysis among the giant panda, the other Ursidae species and lesser panda should be carried out in the future, The CT and image analyzing systems are useful and indispensable for the functional morphology at least in the skeletal system.

Acknowledgements

We wish to express our gratitude to the staff of Kobe Municipal Oji Zoo and Ueno Zoological Gardens for a rare carcass and specimen of the lesser and giant panda. This study was financially supported by The Grant for the Biodiversity Research of the 21st Century COE (A14), Grant-in-Aids for Scientific Research nos. 17405018, 18637001 and 19405001 from the Ministry of Education, Science and Culture, Japan. At the IMC-9 (July 31–August 5 2005 at the 9th International Mammalogical Congress in Sapporo, Hokkaido, Japan), the related techniques of the three-dimensional image analysis were presented in a symposium organized by Roland Frey. The JSPS core-to-core program HOPE is one of the strongest financial supporters in this field of research including our symposium at IMC-9.

References

Beijing Zoo, Beijing University, Beijing Agricultural University, Beijing Second Medical College, Beijing Natural History Museum, Shaanxi Zoology Institute (1986). Systematic anatomy and organ-histology. In: Morphology of the Giant panda. Science Press, Beijing, pp 148–152 (in Chinese)

Davis DD (1964) The Giant panda. A morphological study of evolutionary mechanisms. In: Davis DD (ed) Fieldiana zoology memoirs, Vol. 3. Chicago Natural History Museum, Chicago, IL, pp 146–198, pp 41–124

Delisle I, Strobeck C (2005) A phylogeny of the Caniformia (order Carnivora) based on 12 complete protein-coding mitochondrial genes. Molec Phylogenet Evol 37:192–201

Endo H, Sasaki N, Yamagiwa D, Uetake Y, Kurohmaru M, Hayashi Y (1996) Functional anatomy of the radial sesamoid bone in the giant panda (*Ailuropoda melanoleuca*). J Anat 189:587–592

Endo H, Yamagiwa D, Hayashi Y, Koie H, Yamaya Y, Kimura J (1999a) Role of the giant panda's 'pseudo-thumb'. Nature 397:309–310

Endo H, Makita T, Sasaki M, Arishima K, Yamamoto M, Hayashi Y (1999b) Comparative anatomy of the radial sesamoid bone in the polar bear (*Ursus maritimus*), the brown bear (*Ursus arctos*) and the Giant panda (*Ailuropoda melanoleuca*). J Vet Med Sci 61:903–907

Endo H, Hayashi Y, Yamagiwa D, Kurohamaru M, Koie H, Yamaya Y, Kimura J (1999c) CT Examination of the manipulation system in the giant panda (*Ailuropoda melanoleuca*). J Anat 195:295–300

Endo H, Sasaki M, Hayashi Y, Koie H, Yamaya Y, Kimura J (2001a) Carpal bone movements in gripping action of the giant panda (*Ailuropoda melanoleuca*). J Anat 198:243–246

Endo H, Sasaki M, Kogiku H, Yamamoto M, Arishima K (2001b) Radial sesamoid bone as a part of the manipulation system in the lesser panda (*Ailurus fulgens*). Ann Anat 183:181–184

Endo H, Sasaki M, Narushima E, Komiya T, Hayashida A, Hayashi Y, Stafford BJ (2003a) Macroscopic study of the

functional significance of the forearm muscles in the giant panda. J Vet Med Sci 65:839–843
Endo H, Watanabe K, Hayashi Y, Matsuzaki M, Sakai T, Itou T, Koie H, Kimura J, Stafford BJ (2003b) Three-dimensional CT image analysis of the digging system in the aardvark. Ann Anat 185:367–372
Endo H, Hama N, Niizawa N, Kimura J, Itou T, Koie H, Sakai T (2007a) Three-dimensional reconstruction of the manipulation system in the lesser panda. Mammal Study 32:99–103
Endo H, Niizawa N, Komiya T, Kawada S, Kimura J, Itou T, Koie H, Sakai T (2007b) Three-dimensional CT examinations of the mastication system in the giant anteater. Zool Sci 24:1005–1011
Fulton TL, Strobeck C (2006) Molecular phylogeny of the Arctoidea (Carnivora): Effect of missing data on supertree and supermatrix analyses of multiple gene data sets. Molec Phylogenet Evol 41:165–181
Gould SJ (1978) The panda's peculiar thumb. Natural History 87:20–30
Pocock RI (1939) The prehensile paw of the giant panda. Nature 143:206
Wood-Jones F (1939a) The forearm and manus of the giant panda, Ailuropoda melanoleuca, M.-Edw. with an account of the mechanism of its grasp. Proc Zool Soc Lond Ser B 109:113–129
Wood-Jones F (1939b) The 'thumb' of the giant panda. Nature 143:157

6 Using CT to Peer into the Past: 3D Visualization of the Brain and Ear Regions of Birds, Crocodiles, and Nonavian Dinosaurs

Lawrence M. Witmer[1], Ryan C. Ridgely[1], David L. Dufeau[2], and Molly C. Semones[3]

6.1 Introduction

Until relatively recently, information on the internal skull structures of fossil taxa relied on fortuitous breaks, aggressive removal of rock matrix (Galton 1989, 2001), sectioning with a saw (Osborn 1912), or serial ground thin-sectioning (Stensiö 1963), all of which potentially risk damage to the fossil specimen (or even consume it entirely in the case of ground thin-sections). In some cases, casts of internal structures, such as the brain cavity and labyrinth of the inner ear, were preserved as 'natural endocasts' by infilling with more resistant matrix (e.g., Newton 1888). In most cases, however, physical endocasts are made after matrix removal by coating internal cavities with latex and then removing the cured replica, referred to as a latex endocast (Radinsky 1968; Jerison 1973; Hopson 1979). The process of making latex endocasts poses further risks to the fossil, and for many fragile specimens, such an approach has been unfeasible.

The advent of non-invasive, non-destructive radiological approaches – specifically, X-ray computed tomography, better known as CT or CAT scanning – has revolutionized the study of fossil specimens (Conroy and Vannier 1984; Carlson et al. 2003), allowing the investigator to see through the bone and rock matrix and visualize internal structures. Moreover, CT scanning is an effective 3D digitizing tool, mapping all points within the subject into x-y-z coordinate space. This property allows features of interest (such as the brain cavity, inner ear, neurovascular canals, air sinuses, etc.) to be digitally highlighted and extracted for visualization and quantification (e.g., Sampson and Witmer 2007). In particular, CT scanning has rejuvenated the subject of 'paleoneurology,' the study of brain evolution in the fossil record. Paleoneurology previously relied exclusively on physical (natural or latex) endocasts (Jerison 1973; Edinger 1975; Hopson 1979; Hurlburt 1996; Buchholtz and Seyfarth 1999; Wharton 2002). Recent years, however, have witnessed an explosion of CT-based studies of the brain cavity (and sometimes also the labyrinth) in a range of vertebrates (Rogers 1998, 1999, 2005; Brochu 2000, 2003; Larsson 2001; Marino et al. 2003; Domínguez Alonso et al. 2004; Franzosa 2004; Franzosa and Rowe 2005; Maisey 2005; Kurochkin et al. 2006; Macrini et al. 2006; Kundrát 2007).

We present some results here of our large CT-based project on the evolution of the brain and ear regions of archosaurs (Witmer et al. 2003; Holliday et al. 2006; Sampson and Witmer 2007; Witmer and Ridgely, in press; Hurlburt et al. in press). Archosaurs are the group of animals that includes birds and crocodilians today and in the Mesozoic Era included nonavian dinosaurs, pterosaurs, and a variety of early forms. The project has sampled taxa from throughout Archosauriformes, generating to date about 150 CT datasets of extinct

[1]Department of Biomedical Sciences and College of Osteopathic Medicine, [2]Department of Biological Sciences, [3]Honors Tutorial College, Ohio University, Athens, Ohio 45701 USA

taxa. Moreover, there has likewise been an extensive sampling of extant taxa, and not just archosaurs, but also the extant outgroups of archosaurs (lepidosaurs, turtles, and mammals). This study is framed within the extant phylogenetic bracket approach (Witmer 1995) whereby the extant outgroups (e.g., birds, crocodilians, lizards) of the fossil taxon of interest (e.g., dinosaurs) provide critical information on the causal associations between soft tissues and their osteological correlates. That is, the extant taxa and their still-present soft tissues provide essential guides to the significance of particular bony details, which can then be directly compared with the bony structures of the fossil specimens. Thus, although the study presented here focuses on certain dinosaur taxa, the extant taxa (Australian freshwater crocodiles and great horned owls) provide the necessary groundtruthing to interpret and infer the soft-tissue structures of dinosaurs (Fig. 6.1).

The dinosaurs we present here include two members of the long-necked sauropod clade, *Camarasaurus lentus* and *Diplodocus longus*. These two sauropods not only allow comparison of variation in brain and ear structure within a group of dinosaurs, but also demonstrate differences in head orientation as indicated by labyrinth orientation (Witmer et al. 2003). Sauropod brain and ear structure will then be compared to that of the gigantic predatory theropod dinosaur *Tyrannosaurus rex* to provide some insight into the sensorineural differences between carnivorous and herbivorous dinosaurs. Rather than provide elaborate anatomical descriptions in the text, we will draw on the benefits of advanced 3D visualization, and let our illustrations 'do the talking.' The text will emphasize comparison and interpretation.

Institutional abbreviations – AMNH, American Museum of Natural History, New York City,

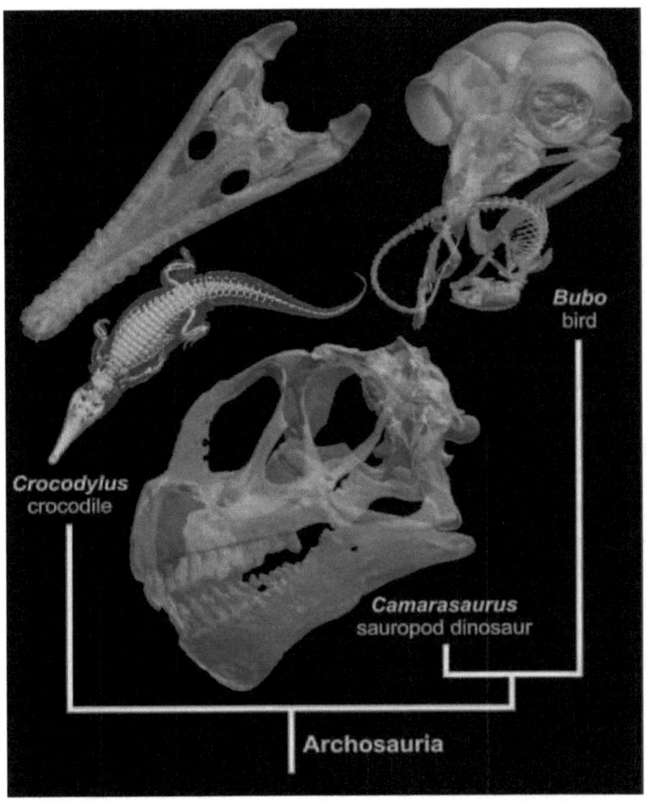

FIG. 6.1. The extant phylogenetic bracket (Witmer 1995) of nonavian dinosaurs, as exemplified by the sauropod dinosaur *Camarasaurus lentus* (CM 11338, in left lateral view). The extant outgroups of the fossil taxa provide information on attributes such as soft-tissue morphology and behavior that are not preserved in the fossils. The extant outgroups of nonavian dinosaurs are crocodilians (as exemplified by *Crocodylus johnstoni*, OUVC 10425, in dorsal view) and birds (as exemplified by *Bubo virginianus*, OUVC 10220, in left rostrolateral view with *Mus musculus* [OUVC 10449] in its jaws). The images are surfaces renderings of CT scan data, with the skull rendered semitransparent revealing the cranial endocast (*in blue*) and other structures (see Color Plates, Fig. 6.1).

New York; ANSP, Academy of Natural Sciences, Philadelphia, Pennsylvania; CM, Carnegie Museum of Natural History, Pittsburgh, Pennsylvania; HMS, Houston Museum of Natural Science, Houston, Texas; ISIR, Indian Statistical Institute, Calcutta, India; OUVC, Ohio University Vertebrate Collections, Athens, Ohio.

Anatomical Abbreviations – as, articular sinus; bsr, basisphenoid recess(es); c, cochlea; car, cerebral carotid artery canal; cbl, cerebellum; cer, cerebral hemisphere; col, columella (= stapes); crc, crus communis; csc, caudal (posterior vertical) semicircular canal; csca, ampulla of caudal semicircular canal; ctp, cavum tympanicum proprium; ctr, caudal tympanic recess; cvcm, caudal middle cerebral vein; de, dural expansion; dp, dural peak; dls, dorsal longitudinal sinus; dtr, dorsal tympanic recess; dv, diploic vein draining bone tissue; ed, endolymphatic duct; ev, ethmoid vein; fc, fenestra cochleae (= round window); fl, flocculus (= cerebellar auricle); fp, foramen perilymphaticum (sauropods); fv, fenestra vestibuli (= oval window); h, hyperpallium (= Wulst, eminentia sagittalis); itr, intertympanic recess; lab, endosseous labyrinth; lsc, lateral (horizontal) semicircular canal; lsca, ampulla of lateral semicircular canal; lscr, lateral subcondylar recess; mpr, median pharyngeal recess; mscr, medial subcondylar recess; ob, olfactory bulb; ocv, orbitocerebral vein; oevc, orbital emissary vein canal; opt, optic tectum (= lobe); ovs, occipital venous sinus; pfo, pituitary (= hypophyseal) fossa; pin, pineal peak; po, pons; ptr, pharyngotympanic (= Eustachian) recess; qs, quadrate sinus; rde, rostral dural expansion; rsc, rostral (anterior vertical) semicircular canal; rsca, ampulla of rostral semicircular canal; rtr, rostral tympanic recess; rvcm, rostral middle cerebral vein; s, siphonium; sin, blind dural venous sinus of hindbrain; sps, sphenoparietal sinus; ssr, subsellar recess; st, stapedial artery; ts, transverse sinus; tvm, tuber ventromediale of the cerebrum; vc, venous canal; vcd, dorsal head vein; ve, vestibule of inner ear; vls, ventral longitudinal sinus; vt, vallecula telencephali; I, olfactory canal; II, optic nerve canal; III, oculomotor nerve canal; IV, trochlear nerve canal; V_1, ophthalmic nerve canal; V_2, maxillary nerve canal; V_{2-3}, maxillomandibular nerve canal; V_{2-3}/VII, common external opening in braincase for maxillomandibular and facial nerve canals; $V_{2\text{-so}}$, canal for supraorbital branch of maxillary nerve; V_3, mandibular nerve canal; V_{gang}, trigeminal (Gasserian) ganglion; V_{tym}, tympanic branch of trigeminal nerve canal; VI, abducens nerve canal; VII, facial nerve canal; VII_h, canal for hyomandibular ramus of facial nerve; VII_{pal}, canal for palatine ramus of facial nerve; IX, glossopharyngeal nerve canal; IX–XI, shared canal for glossopharyngeal, vagus, and accessory nerves and accompanying vessels; X, vagal nerve canal; X_{tym}, canal for tympanic branch of glossopharyngeal and vagus nerves; XII, hypoglossal nerve canal.

6.2 Materials

As noted above, the broader sample of extinct archosaurs and extant amniotes that have been analyzed includes many dozens of taxa. The extant archosaur taxa presented here are the following: (1) Crocodylia, Crocodylidae, *Crocodylus johnstoni* (Australian freshwater crocodile; OUVC 10425, an intact frozen carcass of an adult captive individual); and (2) Aves, Strigiformes, Strigidae, *Bubo virginianus* (great horned owl; OUVC 10220, a dried skull of an adult wild individual). The fossil dinosaur taxa presented here are the following: (1) Dinosauria, Saurischia, Sauropodomorpha, Macronaria, *Camarasaurus lentus* (CM 11338, a nearly complete skull of a juvenile individual); (2) Dinosauria, Saurischia, Sauropodomorpha, Diplodocoidea, *Diplodocus longus* (CM 11161, a nearly complete skull of an adult individual; CM 3452, a nearly complete skull of a subadult individual; AMNH 694, a sagittally sectioned braincase of a subadult individual); and (3) Dinosauria, Saurischia, Theropoda, Coelurosauria, *Tyrannosaurus rex* (AMNH 5117, a braincase of a subadult individual). Additional specimens of *T. rex* and other tyrannosaurids also were studied in connection with other projects (see Witmer and Ridgely in press for details). All of the sauropod specimens were collected from the Late Jurassic Morrison Formation; CM 11338, 11161, and 3452 come from Dinosaur National Monument, Utah, and AMNH 694 comes from Bone Cabin Quarry, Wyoming. AMNH 5117 was collected from the Late Cretaceous Lance Formation of Converse County, Wyoming. Figure 6.2 presents a cladogram depicting the relationships of these five focal taxa.

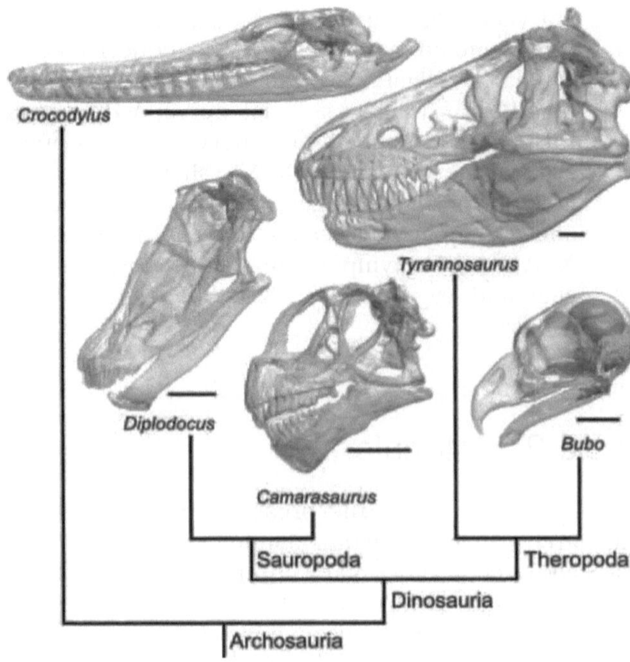

FIG. 6.2. Diagram of the phylogenetic relationships of the principal taxa discussed in the text. From left, the extant crocodilian *Crocodylus johnstoni* (OUVC 10425), the diplodocoid sauropod *Diplodocus longus* (CM 11161), the macronarian sauropod *Camarasaurus lentus* (CM 11338), the basal coelurosaurian theropod *Tyrannosaurus rex* (AMNH 5117), and the avian theropod (bird) *Bubo virginianus* (OUVC 10220). These sauropods and theropods are saurischian dinosaurs. The images are surface renderings of CT scan data. Bony skulls are depicted as transparent, revealing such internal structures as the brain cast, labyrinth, and pneumatic sinuses. All skulls are oriented in their 'alert' postures, determined by orienting the skull such that the lateral semicircular canal is horizontal. Scale bars equal 10 cm, except that for *B. virginianus* which equals 2 cm (see Color Plates, Fig. 6.2).

6.3 Methods

CT scanning of the extant crocodile (OUVC 10425), *Camarasaurus* (CM 11338), *Diplodocus* (CM 3452), and *Tyrannosaurus* (AMNH 5117) took place at O'Bleness Memorial Hospital (OBMH) in Athens, Ohio, using a General Electric (GE) LightSpeed Ultra Multislice CT scanner. This scanner was equipped with the Extended Hounsfield option (which greatly improves resolvability of detail from dense objects such as fossils by extending the dynamic range of images as much as 16-fold) and a 'bow-tie' filter (which decreases beam-hardening artifacts). These specimens were scanned helically at a slice thickness of 625 μm, 120 kV, and 200–300 mA. The other specimens of *Diplodocus* were scanned at OBMH with a GE HiSpeed FX/i Multislice helical scanner at a slice thickness of 1 mm (AMNH 694) or 2 mm (CM 11161), 120–140 kV, and 150 mA. The raw scan data were reconstructed using a bone algorithm. Data were output from the scanners in DICOM format, and then imported into Amira 3.1.1 or 4.1.1 (Mercury-TGS, Chelmsford, MA) for viewing, analysis, and visualization.

The extant owl skull (OUVC 10220) was scanned at the Ohio University MicroCT (OUμCT) facility in Athens, Ohio, using a GE eXplore Locus in vivo Small Animal MicroCT Scanner. It was scanned (with a domestic mouse, *Mus musculus* [OUVC 10449], in its jaws; Fig. 6.1) at a slice thickness of 92 μm, 80 kV, 450 μA. The resulting volume data (in VFF format) were exported from MicroView 2.1.2 (open-source software developed by GE; microview.sourceforge.net) in DICOM format, which were then subsequently imported into Amira.

Data derived from both OBMH and OUμCT were analyzed on 32- and 64-bit PC workstations with 4 GB of RAM and nVidia Quadro FX 3000 or 4500 video cards and running Microsoft Windows XP Professional, Windows XP Professional x64, or Linux 2.6.18 (Debian 4.0 distribution). Structures of interest (e.g., cranial endocast, labyrinth, air sinuses, etc.) were highlighted and digitally extracted using Amira's segmentation tools for quantification and visualization. Both surfaces and volumes were generated, and were used to illustrate this paper. To facilitate discussion, we will refer to the digital casts of structures as if they were the structures themselves (e.g., 'lateral semicircular canal' versus 'digital cast of lateral semicircular canal'). Additional visualizations of the CT scan data are available on www.ohio.edu/WitmerLab.

6.4 Results

As noted earlier, we will rely more on illustration than text to convey anatomical form. We begin this section with a presentation of the extant taxa because the identifications of soft-tissue structures, as well as their relationships to bony structures, have been confirmed by other means (e.g., dissection). Moreover, in the case of the crocodile, an intact specimen was scanned, and many soft-tissue components can be observed in the scan data. We follow with a presentation of the fossil taxa, beginning with the two sauropods and then *Tyrannosaurus*, basing our identifications in part on the crocodile and owl presented here but also on the dozens of other extant and fossil archosaurs we have analyzed as part of the broader project.

6.5 Extant Outgroups: *Crocodylus johnstoni*, Australian Freshwater Crocodile

The cranial endocast, endosseous labyrinth, and columella of *Crocodylus johnstoni* (OUVC 10425) are depicted in Figure 6.3. The endocast resembles in many ways those generated for *C. acutus* (Colbert 1946a) and *C. moreletii* (Franzosa 2004), although this is the most detailed cranial endocast published to date for any crocodilian. As is generally the case for reptiles (Jerison 1973; Hopson 1979), the brain itself does not fill the endocranial cavity, and thus the endocast is in large measure a cast of the dural envelope. As a result, many of the brain parts are obscured by the overlying dural venous sinuses and the dura itself. For example, as

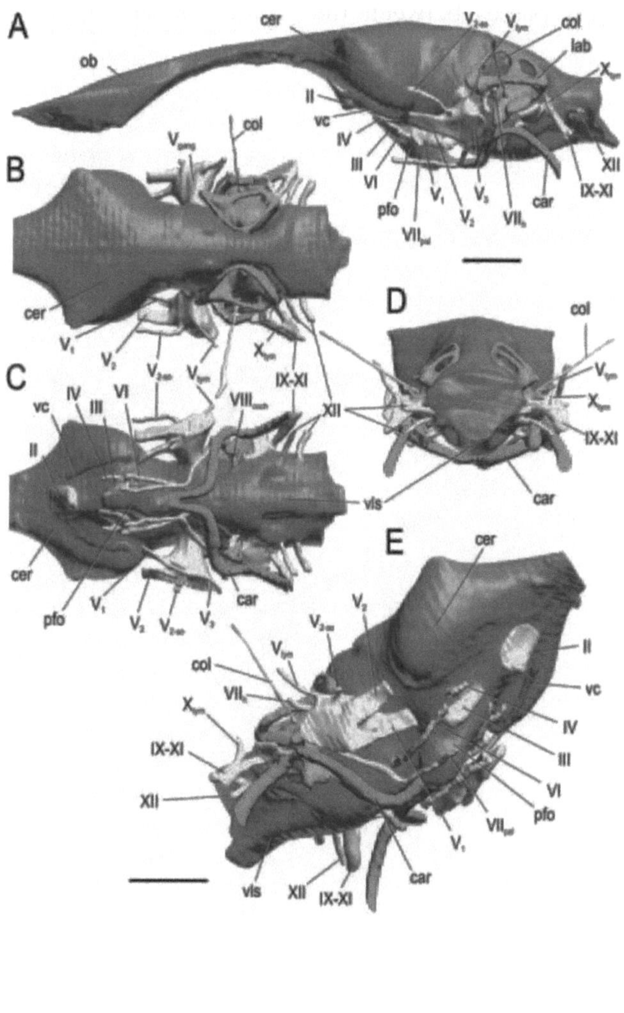

FIG. 6.3. Cranial endocast, endosseous labyrinth, and some endocranial vascular structures of an Australian freshwater crocodile, *Crocodylus johnstoni* (OUVC 10425), derived from surface renderings of CT scan data. **A**, left lateral view. **B**, dorsal view, with olfactory tract truncated. **C**, ventral view, with olfactory tract truncated. **D**, caudal view. **E**, right rostroventrolateral view. Color scheme: cranial endocast, *blue*; endosseous labyrinth, *pink*; nerve canals (most of which also transmit veins), *yellow*; smaller venous canals, *dark blue*; arterial canals, *red*; columella, *pale yellow*. Scale bars equal 1 cm (see Color Plates, Fig. 6.3).

is typical for archosaurs, the large occipital venous sinus (an expansion of the dorsal longitudinal sinus that drains much of the brain and exits through the foramen magnum) overlies and hence obscures the details of the cerebellar, optic tectal, and brain stem regions (Sedlmayr 2002). Likewise the large ventral longitudinal sinus obscures the brainstem ventrally. Nevertheless, telencephalic portions of the brain, such as the olfactory bulbs and cerebral hemispheres, sufficiently fill their regions of the endocranial cavity that the endocast is a more faithful representation of their general size and shape. Although a vascular injection of the specimen at hand was not done, air fortuitously had entered the encephalic venous system post-mortem, providing the opportunity to digitally segment and visualize components of the dural sinus system that discriminate brain regions (Fig. 6.4). Remarkably, the pattern serendipitously recovered closely reflects the published accounts of Hochstetter (1906); van Gelderen (1924) and Hopson (1979). For example, the transverse sinus passes between the regions of the optic tectum and otic capsule, and the sphenoparietal sinus passes between the cerebral hemisphere and tectum.

Figure 6.3 also illustrates the major neurovascular features of the endocast, such as the canals for the cranial nerves and cerebral carotid artery. It should be noted that virtually all of these canals also transmit veins from adjacent regions into the dural venous sinuses (Sedlmayr 2002). Structures of particular note include the facial nerve (CN VII), which splits into its hyomandibular ramus (chorda tympani) and its long palatine ramus. The trigeminal (Gasserian) ganglion is located outside the cranial cavity (i.e., extracranial), and its main branches – the ophthalmic (CN V_1), maxillary (CN V_2), and mandibular (CN V_3) nerves – are well-visualized, as are two lesser known branches: (1) the supraorbital ramus of the maxillary nerve (Holliday and Witmer 2007) and (2) the tympanic branch, which passes through the prootic bone to reach the tympanic cavity. This latter branch relates to another seldom-seen structure that is visible on the endocast, and that is the canal for

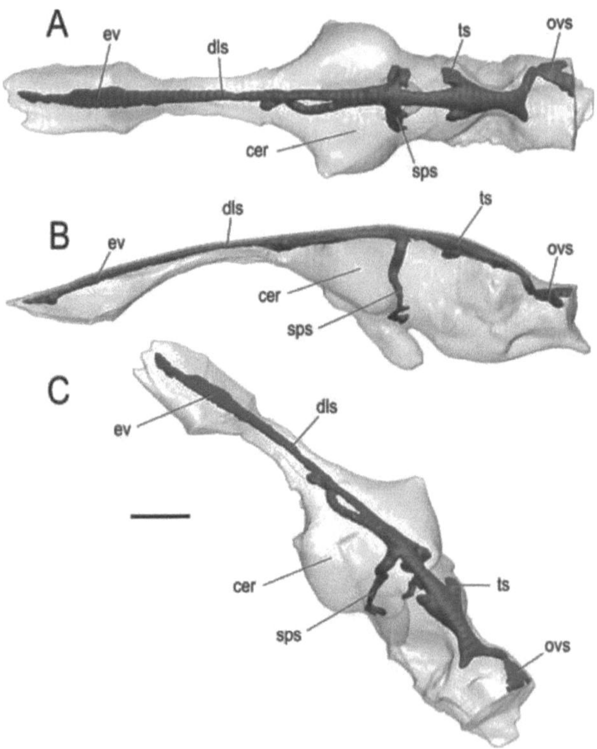

FIG. 6.4. Dural venous sinuses within the semitransparent cranial endocast of an Australian freshwater crocodile, *Crocodylus johnstoni* (OUVC 10425), derived from surface renderings of CT scan data. **A**, dorsal view. **B**, left lateral view. **C**, left caudodorsolateral view. Scale bar equals 1 cm.

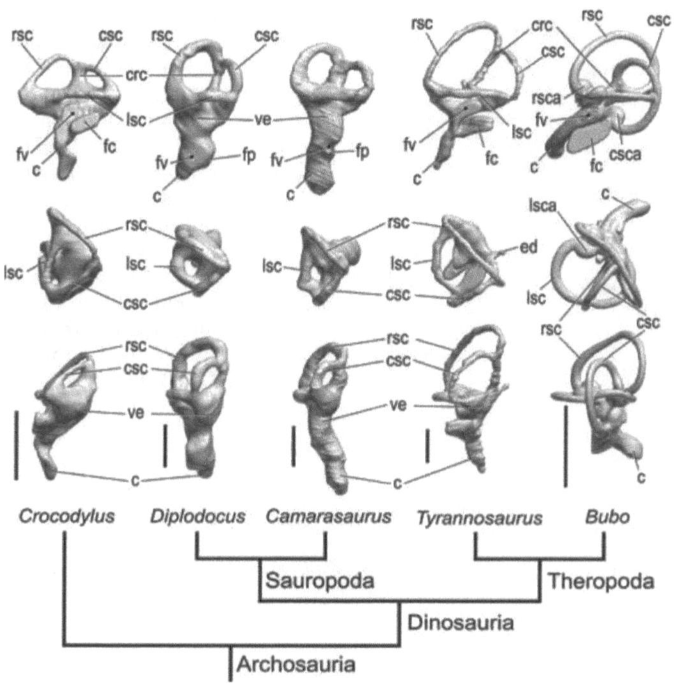

FIG. 6.5. Endosseous labyrinths of the left inner ears of the principal taxa discussed in the text, displayed on a diagram of their phylogenetic relationships. From left, the extant crocodilian *Crocodylus johnstoni* (OUVC 10425), the diplodocoid sauropod *Diplodocus longus* (CM 11161), the macronarian sauropod *Camarasaurus lentus* (CM 11338), the basal coelurosaurian theropod *Tyrannosaurus rex* (AMNH 5117), and the avian theropod (bird) *Bubo virginianus* (OUVC 10220). These sauropods and theropods are saurischian dinosaurs. The images are surface renderings of CT scan data. Top row: left lateral view. Middle row: dorsal view. Bottom row: caudal view. Orientations determined based on orientation of the labyrinth within the skull and with the lateral semicircular canal placed horizontally. Scale bars equal 1 cm.

the tympanic branches of the glossopharyngeal (CN IX) and vagus (CN X) nerves. Killian (1890) regarded these branches as joining with the tympanic branch of the trigeminal nerve mentioned above.

Just as the cranial endocast is a representation of dural form and not strictly the brain itself, the labyrinth does not record the membranous (endolymphatic) labyrinth and moreover does not truly record the 'osseous' labyrinth which traditionally includes the bony canals and walls; rather, it records the space just internal to the bone, and hence we apply the descriptor 'endosseous.' The endosseous labyrinth of the inner ear (Fig. 6.5) is fairly typical for extant crocodilians (we have comparable data for *Alligator*, other *Crocodylus*, *Gavialis*, and numerous fossil species), with a relatively simple, triangular vestibular apparatus dorsally and an elongate cochlea (lagena) ventrally. The strongly twisted cochlea no doubt corresponds to the twisting of the basilar membrane described by Wever (1978). The cerebral carotid artery passes just lateral to the inflection point of the twist. The long slender columellae (stapes) also were visualized (Fig. 6.3) and mark the position of the fenestra vestibuli (ovalis) on the labyrinth. The bony braincase is pneumatized by air-filled diverticula of the pharynx, including a median pharyngeal diverticulum and paired pharyngotympanic (Eustachian) tubes (Fig. 6.6A,B). The general pattern determined from CT is generally similar to that presented by Owen (1850) and Colbert (1946b), although the one presented here is more complete in including the long siphonial tube that passes through the quadrate bone on its way to the articular bone of the lower jaw. Based on broader studies underway by D. L. Dufeau, the cranial pneumatic system of *C. johnstoni* is relatively reduced in comparison to other extant crocodilians.

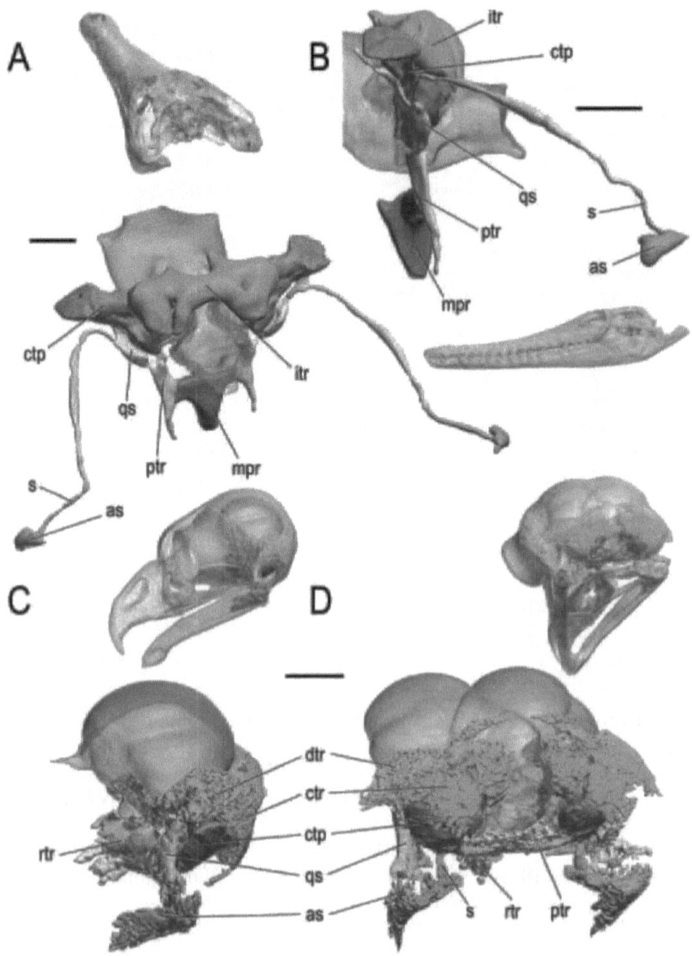

FIG. 6.6. Pneumatic sinuses of the braincase region of, (**A–B**), an Australian freshwater crocodile, *Crocodylus johnstoni* (OUVC 10425), and (**C–D**), a great horned owl, *Bubo virginianus* (OUVC 10220), viewed surrounding a semitransparent cranial endocast, derived from surface renderings of CT scan data. Thumbnail views of the whole skull are provided above the main image (except B, where it is below) to show the orientation of the main image. (**A**) *C. johnstoni* in left caudodorsolateral view. (**B**) *C. johnstoni* in left lateral view. (**C**) *B. virginianus* in left lateral view. (**D**) *B. virginianus* in left caudoventrolateral view. Scale bars equal 1 cm (see Color Plates, Fig. 6.6).

6.6 Extant Outgroups: *Bubo virginianus*, Great Horned Owl

Whereas the cranial endocast of the freshwater crocodile above bears only a superficial resemblance to the underlying brain, the endocast of *Bubo virginianus* (OUVC 10220; Fig. 6.7) closely resembles the form of the brain itself, as determined both by our dissection of a different specimen of *B. virginianus* (OUVC 10427) and Turner's (1891) illustration of a *B. virginianus* brain (see also Stingelin 1957, for photographs of the brains of four other owl species). This close resemblance results from the fact that avian meninges are very thin, and thus the large brain essentially fills the endocranial cavity (Jerison 1973; Iwaniuk and Nelson 2002). A major difference between the brain and endocast is that the expansive occipital dural sinus obscures the cerebellar folia in the endocast. Major features of the endocast include (1) large cerebral hemispheres with a very large hyperpallium (Wulst), (2) small olfactory bulbs, (3) ventrolaterally displaced optic (mesencephalic) tecta (optic lobes), (4) small flocculi (cerebellar auricles), and (5) separate ophthalmic and maxillomandibular branches of the trigeminal nerve reflecting the intracranial position of the trigeminal ganglion. Avian brain terminology used here follows Breazile and Kuenzel (1993); Dubbeldam (1998), and Reiner et al. (2004).

The endosseous labyrinth of the inner ear is a very elongate and elaborate series of canals and swellings (Fig. 6.5), and, as with the brain, much more closely approximates the form of the underlying membranous labyrinth than does the endosseous labyrinth of crocodilians. The remarkable hearing capabilities of owls are well known, and

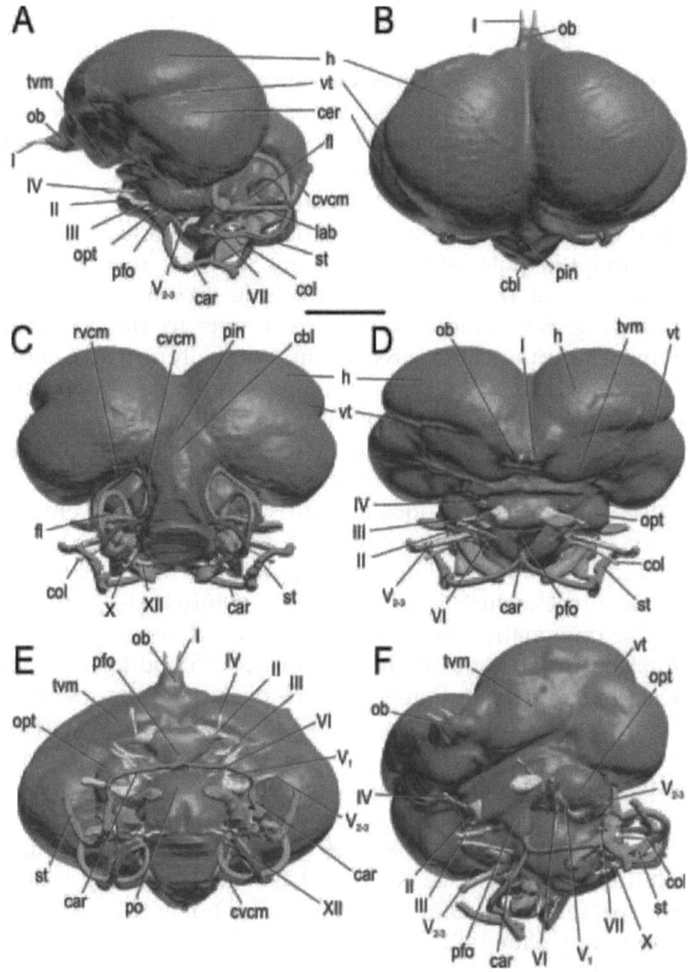

FIG. 6.7. Cranial endocast, endosseous labyrinth, and some endocranial vascular structures of a great horned owl, *Bubo virginianus* (OUVC 10220), derived from surface renderings of CT scan data. **A**, left lateral view. **B**, dorsal view. **C**, caudal view. **D**, rostral view. **E**, ventral view. **F**, left rostroventrolateral view. Color scheme: cranial endocast, *blue*; endosseous labyrinth, *pink*; nerve canals (most of which also transmit veins), *yellow*; smaller venous canals, *dark blue*; arterial canals, *red*; columella, *pale yellow*. Scale bar equals 1 cm (see Color Plates, Fig. 6.7).

their auditory physiology (particularly that of the barn owl, *Tyto alba*) has been very well studied (see reviews by Dooling et al. 2000; Gleich and Manley 2000; Gleich et al. 2004). The structure of the labyrinth itself, however, has been poorly documented in owls, apparently being limited to little more than a schematic drawing (Norberg 1978) and photographs of dissected osseous labyrinths (Tanturri 1933; Turkewitsch 1934).

The semicircular canals are very long and slender. As is almost always the case in birds (based on our CT-based studies of numerous other species), the rostral semicircular canal is the longest and prescribes an oval, whereas the caudal and lateral canals are shorter and more circular. The lateral canal communicates with not just the caudal canal but also the rostral canal. As Gray (1908) noted, perilymphatic communication of the lateral and caudal canals is the typical avian condition, but he found the double-communication condition we report in *B. virginianus* to be quite rare (owls were not included in Gray's sample). The additional communication (between lateral and rostral canals) results in part from the relatively short crus communis, but also from the great length and excursion of the lateral and rostral canals that bring them close enough to communicate.

The footplate of the columella fits into the fenestra vestibuli of the skull, which opens to the base (dorsal end) of the cochlea. The cochlea itself is remarkably long and is curved such that it arcs around and ventral to the pontine portion of the endocast (Fig. 6.7C–F). The great length of the cochlea presumably relates to the presence of an elongate basilar membrane. The length of the basilar membrane is not recoverable from our data and has not been reported otherwise for great horned owls, but barn owls have the longest basilar membranes recorded for birds (Gleich and Manley 2000).

The advantage of CT-based studies of the labyrinth is that structures can be easily viewed in relation to other anatomical systems (Fig. 6.7). For example, the elongate rostral semicircular canal is closely appressed to the cerebellum, overlies the optic tectum rostrally, and is partially overlain by the occipital pole of the cerebrum. In fact, the angulation of the rostral canal visible in dorsal view (Fig. 6.5, middle row) corresponds to the point where the canal passes from cerebellum to cerebrum (Fig. 6.7C), and probably has no relevance for vestibular function. The cerebellar flocculus enters the cage formed by the three canals, passing just rostral to the crus communis, but does not fill the cage, as it does in pterosaurs (Witmer et al. 2003). Much of the labyrinth is surrounded by and suspended within the paratympanic pneumatic sinuses, which is fairly typical for birds.

The pneumatic system itself is very extensive, much more so than in the freshwater crocodile described above, and only its more proximal portions are indicated in Fig. 6.6C–D. Unlike the crocodile, the pneumaticity of the braincase derives ultimately from the pharyngotympanic tubes. The five major tympanic diverticula observed in other birds (Witmer 1990, 1997) are present in great horned owls (see Norberg 1978, for another owl species). The quadrate and articular bones are extensively pneumatized. Another difference with crocodilians is that the siphonial tube in birds does not pass through the quadrate on its way to the articular. The caudal tympanic recess is broadly confluent with the very extensive dorsal tympanic recess. The rostral tympanic recess is also expansive and has a broad contralateral communication ventral to the brain cavity. This communication (known as the interaural pathway) has been implicated in the localization of sounds in space in birds generally, although its role in owls has been controversial (see Klump 2000, for a review).

6.7 Sauropod Dinosaurs: *Camarasaurus* and *Diplodocus*

The first presentation of a digital cranial endocast for a sauropod can be credited to Franzosa (2004) in his unpublished doctoral dissertation; his work on a specimen of *Diplodocus* (HMS 175, formerly CM 662) largely agrees with ours. A few other papers, based on traditional physical endocasts, also provide significant descriptions (Osborn 1912; Osborn and Mook 1921; Janensch 1935, 1936; Ostrom and McIntosh 1966; Hopson 1979; Chatterjee and Zheng 2002, 2005; Knoll et al. 2006). The endocasts presented here (Figs 6.8,6.9) are the most complete published to date. Sauropods lacked the extensive paratympanic sinuses that pneumatized the braincases of the crocodile and owl above, and so the discussion will be limited to the endocast and labyrinth.

Sauropod endocasts in general and those of *Camarasaurus* (Fig. 6.8) and *Diplodocus* (Fig. 6.9) in particular are quite distinct from those of extant birds and crocodilians, but they are much more similar to those of crocodilians in that the brain does not fill the endocranial cavity. In fact, the endocasts of both sauropod taxa presented here suggest that the dura was very loose-fitting indeed, in that brain regions are difficult to discern. Part of this difficulty results from the fact that the endocranial region of most sauropods appears to have been highly vascular, with very large venous sinuses that created prominent grooves, recesses, and apertures within the bony endocranial surface. Again, as in crocodilians, the cerebral hemispheres and olfactory bulbs are the clearest, although even here only portions of the brain regions (e.g., the ventral and lateral surfaces of the cerebrum) are discernible. It should be pointed out that not all sauropods have such a loose dural envelope in that the endocasts of some titanosaurs from India (under study in collaboration with S. Chatterjee) have very clearly defined cerebral regions and a generally less complicated endocast.

From what can be discerned about brain structure, the new digital endocasts confirm much of what previously was known, in that sauropod brains were apparently relatively primitive in overall structure and certainly very small in relation to body size (metric data will be presented elsewhere). Our new studies, however, present the first view of the olfactory bulbs of sauropods. Although only the dorsal contour of the olfactory bulb can be observed due to lack of ossification of the flooring cartilages, criteria developed from the study of modern relatives and other dinosaurs specify the demarcation between olfactory bulb and nasal cavity. The bulb region is better preserved in *Diplodocus* (Fig. 6.9D) than in *Camarasaurus* (Fig. 6.8D), but in both cases the olfactory bulbs are relatively large in comparison to the rest of the endocast. As a result of the caudal retraction of the nasal cavity and concomitant telescop-

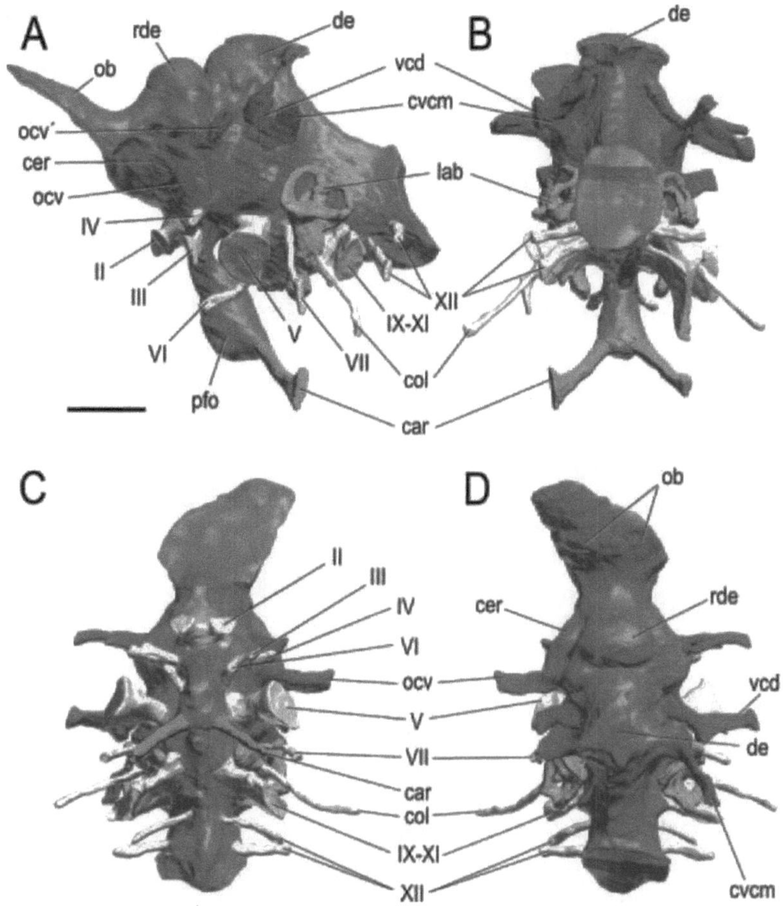

FIG. 6.8. Cranial endocast, endosseous labyrinth, and some endocranial vascular structures of the sauropod dinosaur, *Camarasaurus lentus* (CM 11338), derived from surface renderings of CT scan data. **A**, left lateral view. **B**, caudal view. **C**, ventral view. **D**, dorsal view. Color scheme: cranial endocast, *blue*; endosseous labyrinth, *pink*; nerve canals (most of which also transmit veins), *yellow*; smaller venous canals, *dark blue*; arterial canals, *red*; columella, *pale yellow*. Scale bar equals 2 cm (see Color Plates, Fig. 6.8).

ing of the braincase, the olfactory tracts are quite short, and the bulbs are strongly inclined relative to the main axis of the brain. Another prominent feature of the endocast is the large and pendant pituitary (hypophyseal) fossa. The large bony fossa housed much more than the pituitary itself, in that the cerebral carotids enter it ventrolaterally, the abducens and oculomotor nerves pass through it, and it receives large veins from the orbit (see below). Nevertheless, as Edinger (1942) noted, it is very reasonable to assume that at least part of the reason that these large-bodied dinosaurs had such large pituitary fossae was to house an enlarged pituitary.

The cranial nerve exits on the sauropod endocasts are generally very straightforward, and share a number of common features, many of which are primitive for archosaurs, if not sauropsids as a whole. Shared features, all of which are bilateral, include the following: (1) two canals for the hypoglossal nerve (CN XII); (2) a large 'jugular' aperture (sometimes called the 'metotic foramen;' see discussion in Sampson and Witmer, 2007) through which passed the glossopharyngeal, vagus, and accessory nerves (CN IX–XI) and the jugular vein, and which housed part of the perilymphatic apparatus; (3) a single facial nerve (CN VII) canal; (4) a single canal for the trigeminal nerve which expands laterally as the swelling for the (extracranial) trigeminal ganglion; (5) the abducens nerve (CN VI) canal passes through the lateral portion of the pituitary fossa; (6) the oculomotor (CN III) and trochlear (IV) apertures are large, separate (or sharing an hour-glass-shaped fissure), and located in the infundibular region (where the pituitary fossa joins the main endocranial cavity); and (7) the optic nerve (CN II) has a broad separate aperture. The trigeminal arrangement merits comment in that none of the trigeminal branches are separate in these (or any other known) sauropods, which is unlike the situation in the extant archosaurs discussed above in which the ophthalmic nerve had its own canal. This is discussed further below.

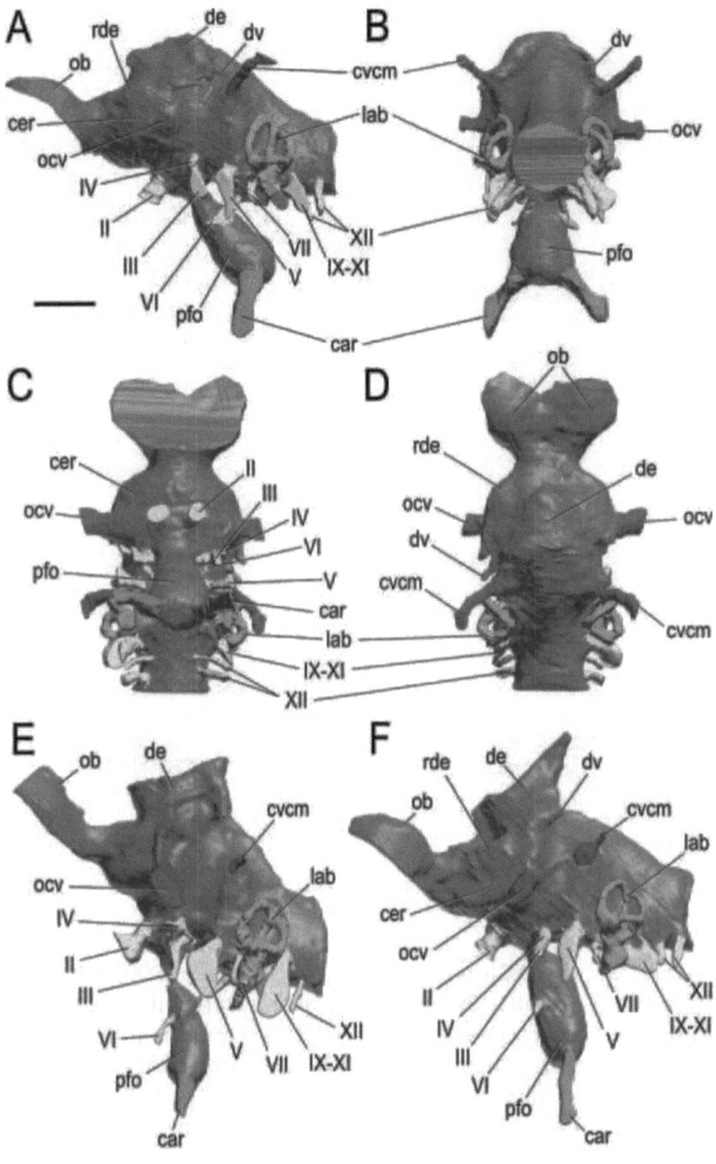

Fig. 6.9. Cranial endocast, endosseous labyrinth, and some endocranial vascular structures of the sauropod dinosaur, *Diplodocus longus*, derived from surface renderings of CT scan data. **A–D**, CM 3452. **E**, AMNH 694. **F**, CM 11161. **A**, **E**, **F**, left lateral view. **B**, caudal view. **C**, ventral view. **D**, dorsal view. Color scheme: cranial endocast, *blue*; endosseous labyrinth, *pink*; nerve canals (most of which also transmit veins), *yellow*; smaller venous canals, *dark blue*; arterial canals, *red*. Scale bar equals 2 cm (see Color Plates, Fig. 6.9).

There has been some confusion in the literature regarding the location of the trochlear nerve aperture. For example, Osborn (1912) identified in *Diplodocus* (AMNH 694) a large aperture between the orbitosphenoid and laterosphenoid as being for the trochlear nerve; the endocast he presented for this specimen has a corresponding swelling. This identification was subsequently followed by Hopson (1979), Galton (1985), and Franzosa (2004). Our findings, on the other hand, suggest that the aperture in question is a large venous channel and that the trochlear foramen is located more ventrally, directly above the oculomotor foramen (Figs 6.8A,C, 6.9A,C). Our digital endocasts of both *Diplodocus* and *Camarasaurus* have three apertures along the laterosphenoid-orbitosphenoid contact: the oculomotor and trochlear foramina together ventrally and the venous channel displaced dorsally. This identification is supported by the fact that the trochlear foramen in some other saurischian dinosaurs (see Sampson and Witmer 2007) is in the same general region as the foramina for the optic, oculomotor, and abducens nerves, which makes sense given their related functions. It should be noted that Janensch (1935, 1936) showed a similar arrangement of foramina for *Barosaurus*, but regarded the trochlear and oculomotor nerves as sharing a large foramen in *Dicraeosaurus*. Likewise, we have found both conditions in our broader sample, with, for example, the Indian titanosaur specimens ISIR 603 and ISIR 199 displaying separate but adjacent trochlear and oculomotor foramina and *Suuwassea emilieae* (ANSP 21122) displaying a common aperture for the two nerves, as in *Dicraeosaurus*.

The identity of the large, more dorsal foramen that formerly had been regarded as being for the trochlear nerve is not clear. Janensch (1935, 1936) regarded it as being homologous to the fenestra epioptica, an opening within the chondrocrania of most extant sauropsids. He further suggested that the aperture transmitted an encephalic vein, the 'anterior cerebral vein.' His assessment was followed by most workers, even those that thought the aperture also transmitted the trochlear nerve (e.g., White 1958; Galton 2001; Franzosa 2004). It is certain that a large vein drained from the orbit into the encephalic sinuses via this foramen, but homology with the embryonic fenestra epioptica of Gaupp (1900) is far from certain, as is the identity of the vein as the 'anterior cerebral vein.' This particular vein traverses the fenestra epioptica for only a brief period in some squamate embryos (van Gelderen 1924), and usually nothing traverses the fenestra epioptica (Shiino 1914). Thus, we instead apply the new term 'orbitocerebral foramen,' which transmitted a vein of the same name (Figs 6.8,6.9). Our CT-based studies have revealed that many groups of dinosaurs have fairly extensive anastomoses between veins within the cranial cavity (i.e., the dural sinuses) and the roof of the orbit (described for the theropod *Majungasaurus* in Sampson and Witmer 2007), and we regard the orbitocerebral foramen of sauropods as transmitting a particularly large such anastomosis.

The large size of the orbitocerebral foramen is just one of several pieces of evidence indicating that the endocranial cavity of many sauropods had an unusually complicated venous system. Virtually all of the cranial nerve foramina, particularly those opening into the orbit, are enlarged – certainly much larger than can be explained by transmitting just the nerve trunks. Given that the cranial nerve foramina in both clades of extant archosaurs (birds and crocodilians) transmit veins into the dural sinus system, it is reasonable to argue that not only did sauropods have such venous anastomoses but also that they were very extensive. Other veins draining into the dural sinuses also appear to have been large. For example, *Camarasaurus* (CM 11338) has an additional orbitocerebral vein aperture on each side (Fig. 6.8A), and both the middle cerebral and dorsal head vein canals are well developed. These venous canals are more modest in specimens of *Diplodocus*, and the dorsal head vein and extra orbitocerebral vein canals are absent (Fig. 6.9).

Perhaps the most curious attribute of the cranial endocasts of most sauropods pertains to the dorsal dural expansion just caudal to the cerebral region. In the *Camarasaurus* and *Diplodocus* specimens at hand, there is a large median expansion with paired ventrolateral communications with venous features. In *Diplodocus*, these venous tracts communicate with the orbitocerebral apertures, whereas in *Camarasaurus* these tracts lead to the dorsal-head/middle-cerebral vein system. *Diplodocus* also has paired dural projections arising from the median dural expansion that extend blindly into the adjacent bone, and presumably represent diploic veins (Fig. 6.9A,B); these are much less developed in *Camarasaurus*. *Camarasaurus*, however, has an additional, median expansion rostral to and smaller than the one just discussed (Fig. 6.8A,D), and Chatterjee and Zheng (2005) illustrated a similar feature in a different specimen. *Diplodocus* also has a rostral dural feature, but here it takes the form of paired swellings that arc toward the olfactory tract cavity above the cerebral region (Fig. 6.9A,D).

The significance and soft-tissue relations of the dorsal dural expansion have been unclear. In many sauropods, this dural expansion is associated with an aperture (known as the parietal fontanelle) between the frontal and parietal bones, suggesting the presence of a pineal organ to some workers (Osborn 1912; Janensch 1935). Although some specimens of *Camarasaurus* have the aperture, CM 11338 lacks it, as reported by Gilmore (1925) and Madsen et al. (1995); we agree, although CT scanning reveals the bone to be very thin in this area. With regard to *Diplodocus*, both CM 11161 and AMNH 694 have large apertures, although the edges are broken, and it is hard to be certain of their veracity; CM 3452 clearly lacks an aperture. Whether or not there is an aperture ultimately is irrelevant to the inference of a pineal organ because extant dinosaurs (birds) and mammals have pineal glands yet lack skull roof apertures, and such seems to be the case for at least some theropods, as well (Sampson and Witmer 2007). Hopson (1979), followed by a number of authors (e.g., Galton 1985, 2001; Wharton 2002), suggested the possibility that the large dorsal expansion was not dural but actually a gap or fontanelle that in life was plugged with persistent cartilage. We regard this hypothesis as falsified by the fact that the frontal and parietal are dermal elements, and thus there was no cartilage to 'persist.' The pineal hypothesis remains valid but obviously the volume of the dorsal expansions exceed that required for a pineal, and, combined with the other evidence for extensive dural sinuses, we regard the venous hypothesis as being the most likely (and does not preclude presence of the pineal).

The endosseous labyrinth of the inner ear has been reconstructed from the CT data for all four sauropod specimens presented here (and also for many other specimens in our broader sample). The labyrinths of *Camarasaurus* (Figs 6.5,6.8) and *Diplodocus* (Figs 6.5,6.9) are similar to each other, and closely resemble the only other sauropod labyrinth in the literature, that of *Brachiosaurus* (Janensch 1935; Clarke 2005). In general, the labyrinth is much more similar to that of the crocodile presented above than the bird, in that, as with the endocast, it provides only a rough approximation of the membranous labyrinth. For example, the ampullae are not at all apparent. Features of interest shared by *Camarasaurus* and *Diplodocus* include (1) a relatively very short lateral canal; (2) an enlarged vestibule; and (3) a relatively short cochlea. The cochlea superficially appears to be long, but the position of the fenestra vestibuli and columellar footplate fixes the vestibulocochlear junction, indicating that the vestibule was expanded ventrally well below the semicircular canal system (Fig. 6.5). Differences between the sauropod taxa in labyrinth structure are few, but the rostral semicircular canal is shorter and rounder in *Camarasaurus*, as opposed to that of *Diplodocus* and *Brachiosaurus*, which is taller and hence more oval. The endosseous labyrinths of the Indian titanosaurs in our sample resemble *Camarasaurus* in this regard. Based on outgroup comparison with the basal sauropodomorphs (e.g., *Massospondylus*, *Plateosaurus*) and theropods in our sample, the *Camarasaurus* rostral canal condition is derived, as are the short lateral canal and enlarged vestibule of sauropods generally.

6.8 Theropod Dinosaurs: *Tyrannosaurus*

The cranial endocast and endosseous labyrinth of the Cretaceous theropod *Tyrannosaurus rex* is presented briefly here (Fig. 6.10) to provide a counterpoint to those of the sauropods. A more detailed study of tyrannosaurs, including multiple specimens of *T. rex*, other tyrannosaurids, and a sampling of other theropods, is presented elsewhere. In general, the endocast still more closely resembles that of extant crocodilians in reflecting the contours of the brain only in the telencephalon. This condition is obviously primitive, but significant nonetheless in that tyrannosaurids are relatively closely related to birds. The cerebral region is enlarged relative to that in more basal theropods (Larsson et al. 2000; Hurlburt et al. in press), and expanded well beyond that observed in sauropods. Although the size of the olfactory lobes was greatly exaggerated in previous reports (e.g., Brochu 2000), the new corrected lobes presented here still are very enlarged in comparison to outgroups, and suggest that olfaction was behaviorally important to tyrannosaurs. With regard to the nerve trunks, the ophthalmic nerve is clearly separated from the maxillomandibular branch of the trigeminal nerve, suggesting that the point of their union – the trigeminal ganglion – is *internal* to the endocranial wall; thus, tyrannosaurids have the derived avian condition of an intracranial trigeminal ganglion rather than the primitive extracranial position exhibited by sauropods and crocodilians. Tyrannosaurs also resemble birds such as the owl in having a well developed middle-cerebral-vein/transverse-sinus system, yet retain the dorsal-head/middle-cerebral vein anastomoses described above for *Camarasaurus* and present in other theropods (Sampson and Witmer 2007).

In comparison to the other endosseous labyrinths discussed here, the tyrannosaur labyrinth (Fig. 6.5) is intermediate between that of the owl and crocodile in that the semicircular canals are long and slender, the ampullar regions are apparent (although not nearly as distinct as in the owl), and the cochlea is elongate. The air sinuses that pneumatize the braincase region of tyrannosaurs are extensive (Fig. 6.11), and share many attributes with those of advanced theropods and birds (Witmer 1997). As in birds, such as the owl, there

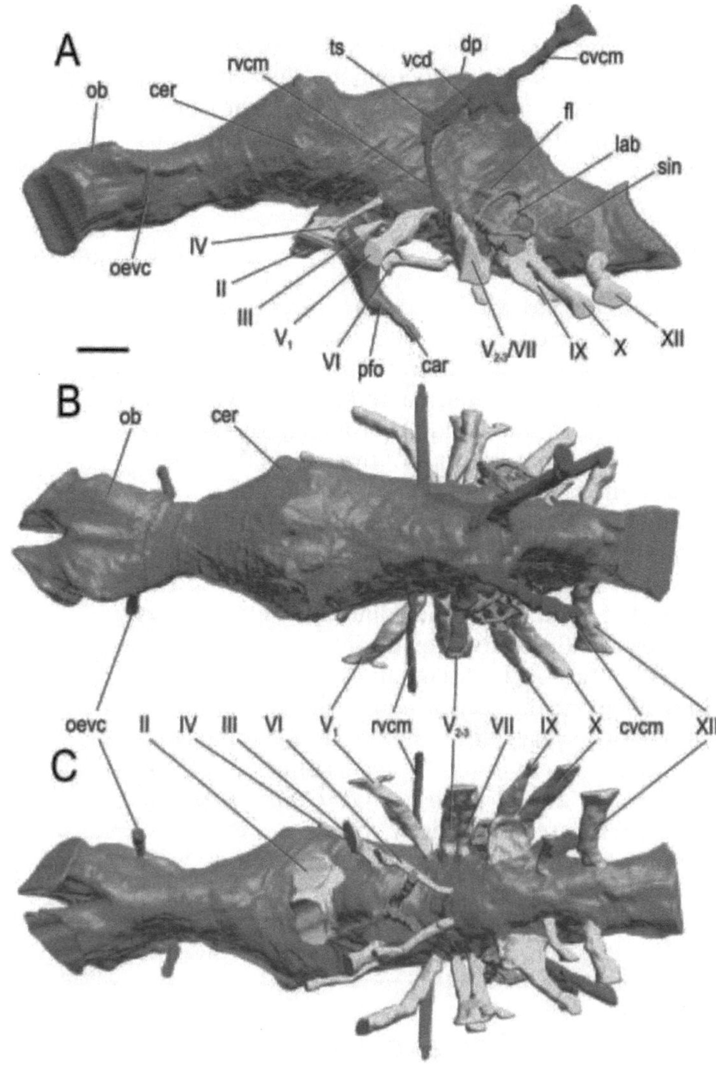

FIG. 6.10. Cranial endocast, endosseous labyrinth, and some endocranial vascular structures of the theropod dinosaur, *Tyrannosaurus rex* (AMNH 5117), derived from surface renderings of CT scan data. **A**, left lateral view. **B**, dorsal view. **C**, ventral view. Color scheme: cranial endocast, *blue*; endosseous labyrinth, *pink*; nerve canals (most of which also transmit veins), *yellow*; smaller venous canals, *dark blue*; arterial canals, *red*. Scale bar equals 2 cm (see Color Plates, Fig. 6.10).

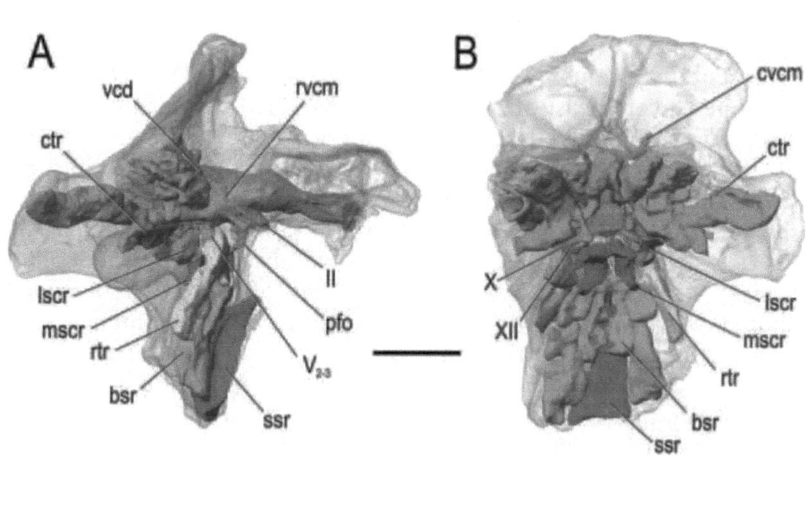

FIG. 6.11. Pneumatic sinuses and cranial endocast within the semi-transparent bony braincase of the theropod dinosaur, *Tyrannosaurus rex* (AMNH 5117), derived from surface renderings of CT scan data. **A**, right lateral view. **B**, caudal view. Scale bar equals 10 cm (see Color Plates, Fig. 6.11).

is a broad communication of the contralateral rostral tympanic recesses below the brain cavity in tyrannosaurs. As in crocodilians, there is broad contralateral communication above the brain cavity, as well, but these are via non-homologous sinuses.

6.9 Discussion

CT scanning indeed provides a means of 'looking inside' an object, which is a useful thing to do whether the object is a modern owl or an ancient dinosaur skull. But the combination of CT with digital segmentation and 3D visualization has provided a new anatomical tool that rivals the invention of the scalpel with regard to scientific impact. In the present case, we have employed these techniques to digitally dissect the anatomical structures of an extant crocodile and bird, as well as sauropod and tyrannosaur dinosaurs. Our figures convey the illustrative power of these approaches, but here we seek to discuss the bearing these findings have for the sensorineural interpretation of these two very different kinds of dinosaurs. Although *Tyrannosaurus* lived some 80 million years after *Camarasaurus* and *Diplodocus* and thus these are not strictly 'predator and prey,' tyrannosaurids and comparable sauropods were sympatric at least in the southwestern United States and in Mongolia (Weishampel et al. 2004), and so the comparison has some validity beyond heuristic value.

6.9.1 Brain Size and Structure

That dinosaurs have small brains relative to body size is hardly a new discovery, and, as mentioned earlier, full quantitative analysis will be presented elsewhere. Nevertheless, we will present some representative values here for comparison. For example, the endocast of *Diplodocus* (CM 11161) has a mass of about 110 g; assuming a body mass of about 13,000 kg (Henderson 1999), endocast size is about 0.001% of body size. For *Tyrannosaurus* (AMNH 5117), the endocast size is about 424 g, which, assuming a body mass of about 5000 kg (Erickson et al. 2004), accounts for about 0.008% of body size. More meaningful comparisons look at allometric scaling of these parameters, and Jerison (1973) proposed the metric of Encephalization Quotient (EQ) which provides a comparison of actual brain size relative to the expected brain size of an animal of that body size. We use Hurlburt's (1996; see also Hurlburt et al. in press) modified equations for reptiles (REQ). Both Jerison and Hurlburt accounted for the fact that the cranial endocast is an overestimate of the true brain size by halving endocast volume to estimate brain volume (i.e., brain size is 50% of endocast size), admittedly a very rough measure, but illuminating nonetheless. The REQ for *Tyrannosaurus* is 2.7, which means that its brain was almost three times the size expected for a reptile of its size. The REQ for *Diplodocus* is 0.41, which means that its brain was less than half the expected size. Thus, and this is no real surprise either, the brain power of the predator exceeds that of the prey. Typically this can be understood to reflect mostly the neural requirements for predation (principally sensory integration), but *Tyrannosaurus* presumably had some cognitive capabilities that sauropods lacked. In fact, the cerebral region of *Tyrannosaurus* is not only more visible in the endocasts than in the sauropods, but it is also measurably expanded over theropod dinosaurs other than the advanced birdlike maniraptorans (Larsson et al. 2000); Hurlburt et al. in press).

In terms of structure, again much of the brain form is hidden beneath the loose dural envelope. Nevertheless, *Tyrannosaurus* and the sauropods both had relatively large olfactory bulbs. As mentioned above, even with the newly corrected size of the olfactory bulbs, *Tyrannosaurus* still has large olfactory bulbs, and this enlargement is a derived feature requiring explanation. Jerison (1973) presented the 'principle of proper mass,' which states that the amount of neural tissue (for us, gross size of a brain region) is proportional to the amount or importance of the functions carried out by those tissues (see Butler and Hodos [2005] for an up-to-date corroboration of this principle). Thus, it is reasonable to assume that the apomorphically large size of the olfactory bulbs in tyrannosaurs suggest that the sense of smell was behaviorally important. The new finding presented here is that at least some sauropods also exhibit large olfactory bulbs, again suggesting the importance of olfactory cues. It remains obscure exactly which behaviors of these dinosaurs required a heightened sense of smell, and there is no reason to believe that a single explanation applies to both groups. The explanation sometimes advanced for *Tyrannosaurus* – namely, habitual if not obligate scavenging (e.g., Horner 1994) – obviously did not apply to sauropods, and in fact, there is ample reason to regard active predation, not pure scavenging, as accounting for the sensorineural adaptations of tyrannosaurs.

6.9.2 Inner Ear: Hearing, Balance, Eye Movements, and Head Posture

There are striking differences between sauropods and tyrannosaurs in the form of the endosseous labyrinth of the inner ear, and these differences relate to a number of different biological traits. The cochlea is the part of the inner ear associated with hearing, and tyrannosaurs have a long cochlea whereas sauropods have a short one. Given that the extant birds and crocodilians in our sample have elongate cochleae (see above), it is tempting to regard the short sauropod cochlea as the derived condition; we are testing that hypothesis in our broader study. The length of the bony cochlea is directly related to the length of the sensory epithelium (the basilar membrane). As a result, cochlear length has been used as a rough proxy for the hearing capabilities of birds and crocodilians (Wever 1978; Gleich and Manley 2000). Thus, it would seem that discrimination of auditory stimuli were more important to tyrannosaurs than to sauropods. Moreover, given the relationships set forth by Gleich et al. (2005), it is reasonable to suggest that tyrannosaurs emphasized, in particular, the reception of low-frequency air-borne sounds. This interpretation is consistent with the extensive middle ear pneumaticity that we report above for tyrannosaurs, because these air sinuses reduce the mechanical stiffness of the middle ear air space at low frequencies (by allowing greater vibrational movement of the columellar apparatus), as well as affecting resonant properties and hence frequency-dependent amplification of sounds (Pickles 1988). Extant birds and crocodilians (such as the owl and crocodile above) also have extensive paratympanic pneumaticity, and experimental studies show that these animals also generally emphasize low frequencies (Dooling et al. 2000). Significantly, sauropods essentially lack middle ear pneumaticity, which, in combination with their short cochleae, supports the notion that fine discrimination of air-borne sounds (particularly low frequencies) was not as important for sauropods.

The upper part of the inner ear – the vestibule and semicircular canals – comprises the vestibular apparatus and is associated with the sense of balance or equilibrium. Again, we see marked morphological differences in these attributes between sauropods and tyrannosaurs: sauropods have short, thick semicircular canals and a large vestibule, whereas tyrannosaurs have elongate, slender canals and a more modest vestibule. Comparison with extant archosaurs shows that the semicircular canals of sauropods are more similar to crocodilians (but perhaps even more reduced) whereas those of *Tyrannosaurus* are more like birds (but not approaching the condition of most birds, and certainly not owls). Semicircular canals sense angular acceleration or turning movements of the head, and increased sizes (length, radius of curvature, arc, etc.) have been linked to such behavioral patterns as enhanced agility and aerobatic/acrobatic ability in birds and primates (Tanturri 1933; Turkewitsch 1934; Spoor and Zonneveld 1998; Spoor 2003; Spoor et al. 2007). It is tempting – and perhaps not unjustified – to suggest that the canal differences between these two dinosaur groups relate to tyrannosaurs being active, agile bipeds and the sauropods being slow-moving quadrupeds. However, how such morphological differences of the vestibular system relate to locomotor modes is somewhat obscure for vertebrates generally (Hullar 2006).

Certainly, even if the links to locomotor behavior are less direct (Graf and Klam 2006), the relationship of labyrinth form to eye movements and gaze-stabilization mechanisms is well established in the context of the vestibulo-ocular and vestibulo-collic reflexes (VOR and VCR; see Spoor 2003; Witmer et al. 2003; Spoor et al. 2007; and references therein). The VOR and VCR involves coordination and compensatory movements of the eyes, head, and neck so that a target moving relative to the subject remains fixed and in-focus on the retina. These mechanisms, and hence the semicircular canal system, are best developed in animals for which tracking movements of the eyes are at a premium, which, in fact, tends to include predatory animals, agile animals, and animals that engage in aerobatic, acrobatic, and rapid arboreal locomotion. In this context, the expanded canals of *Tyrannosaurus* can be interpreted as reflecting an enhanced VOR/VCR in comparison to sauropods and extant crocodilians, and further suggesting that rapid tracking movements of the eyes were important behaviorally.

The apomorphically short, stubby canals of sauropods, on the other hand, particularly the very short lateral canal, must be reflecting decreased importance of or reliance on compensatory movements of the eyes and head. Reduction of the semicircular canals in cetaceans has been linked to reduced eye mobility, eye musculature, and neck mobility (see Spoor and Zonneveld 1998; Spoor 2003; and references therein), but we have no objective evidence to suggest that these explanations pertain to sauropods. Sauropod orbits are large, their scleral ossicles (when preserved, as in CM 11338 and CM 11161) suggest an unreduced eyeball, and their necks are the longest known among vertebrates (although perhaps less flexible than once had been thought; Stevens and Parrish 2005a). Rather it would seem that sauropods simply did not engage in rapid eye, head, and neck movements. This behavioral change is consistent with reduction or even absence of a discernible flocculus in most sauropods, the presence of which has been linked to the VOR and VCR in other archosaurs (see Witmer et al. 2003 and references therein). The strong reduction of the lateral canal further suggests that mediolateral eye and head movements, in particular, were less important, which may be consistent with inferred feeding behaviors involving sagittal (dorsoventral) movements of the head (Barrett and Upchurch 1994).

There is a large body of literature (see Witmer et al. 2003) relating the orientation of the lateral semicircular canal to the 'alert' posture of the head, such that when animals are alert and their senses heightened, they tend to hold their heads in an orientation that places the lateral canal roughly horizontal. Hullar (2006) noted that whereas many of the claims relating labyrinth form to locomotory differences are problematic, the links to head posture are well founded. Our team has demonstrated that alert postures of dinosaurs and other archosaurs (as determined by orienting the lateral canal horizontally in a semitransparent skull in lateral view) were rather variable, even within a clade (Witmer et al. 2003; Sampson and Witmer 2007). Turning to the extant taxa presented in this paper, the bird in our sample, *Bubo virginianus*, has an alert posture that is fairly strongly down-turned, such that the long axis of the skull is declined about 30° below the horizontal (Fig. 6.2). The alert head posture as determined from the lateral canal has not been published previously for any owl, but we can report that the declination angle in a barn owl (*Tyto alba*) in our sample is only about 15°. By contrast, a bald eagle (*Haliaeetus leucocephalus*) in our sample has an almost completely horizontal alert posture (declination angle of about 0°). Likewise, the alert posture of the crocodile in our sample, *Crocodylus johnstoni*, is basically horizontal (Fig. 6.2), and this is true for the other extant crocodilians in our sample (e.g., *C. moreletii*, *Alligator mississippiensis*, *Gavialis gangeticus*). Thus, although birds exhibit a wide range of alert head postures, crocodilians are relatively homogenous.

Turning to the fossil taxa presented in this paper, *Tyrannosaurus rex* has a very slightly down-turned alert posture of the head (5–10°; Fig. 6.2), and that of *Majungasaurus* is also basically horizontal (Sampson and Witmer 2007). However, a number of other theropod dinosaurs have much more strongly down-turned postures, including another tyrannosaurid (*Nanotyrannus*: 25–30°). In the above cited papers, we related these differences in alert posture among theropods to the demands of maintaining a maximal binocular field of view by clearing the snout and/or bony orbital rugosities from the field of view. Similar arguments were also made for differences in head posture between two pterosaurs (Witmer et al. 2003). Of the two sauropods presented here, *Camarasaurus* has a much more horizontal skull, with a declination angle of less than 10°, whereas *Diplodocus* has a very strongly down-turned posture (about 45°, averaging values from CM 3452 and CM 11161; Fig. 6.2). In this case, explanations for these differences may have less to do with binocularity, in part because the very large snout would still obstruct frontal vision in *Camarasaurus*, and in part because, based on first principles, we would not predict a strong selective premium for binocularity in these non-predaceous animals. These differences between these two sauropods, however, do agree with other morphological findings from the skull base and neck, and the differing head postures have been related to differences in feeding strategies (Barrett and Upchurch 1994; Fiorillo 1998; Stevens and Parrish 2005b).

Acknowledgments

We thank Roland Frey and Hideki Endo for their invitation to participate in this volume. We thank Matthew Lamanna and Amy Henrici (CM,

Pittsburgh) for loan of *Camarasaurus* and *Diplodocus* skulls, and Mark Norell and Carl Mehling (AMNH, New York City) for loan of *Diplodocus* and *Tyrannosaurus* skulls. For providing the specimen of *Crocodylus johnstoni*, we thank John Hutchinson (Royal Veterinary College, London), Kent Vliet, David Kledzik, and the St. Augustine Zoological Park, Florida. We have benefited from discussions with many colleagues at many institutions, including in particular Christopher Brochu, Thomas Carr, Sankar Chatterjee, Philip Currie, Jon Franzosa, Peter Galton, Casey Holliday, James Hopson, Grant Hurlburt, Harry Jerison, Timothy Rowe, Jayc Sedlmayr, Paul Sereno, Kent Stevens, and Takanobu Tsuihiji. We also benefited from Tsuihiji's careful reading of a draft of this manuscript. For help with CT scanning, we thank Heather Rockhold, RT, and O'Bleness Memorial Hospital, in Athens, Ohio. For funding, we acknowledge National Science Foundation grants IBN-9601174, IBN-0343744, and IOB-0517257, and the Ohio University College of Osteopathic Medicine.

References

Barrett PM, Upchurch P (1994) Feeding mechanisms in *Diplodocus*. Gaia 10:195–203

Breazile JE, Kuenzel WJ (1993) Systema nervosum centrale. In: Baumel JJ (ed.) Handbook of Avian Anatomy: Nomina Anatomica Avium, 2nd edn. Publications of the Nuttall Ornithological Club No 23, Cambridge, pp 493–554

Buchholtz EA, Seyfarth E-A (1999) The gospel of the fossil brain: Tilly Edinger and the science of paleoneurology. Brain Res Bull 48:351–361

Brochu CA (2000) A digitally rendered endocast for *Tyrannosaurus rex*. J Vertebr Paleontol 20:1–6

Brochu CA (2003) Osteology of *Tyrannosaurus rex*: insights from a nearly complete skeleton and high-resolution computed tomographic analysis of the skull. J Vertebr Paleontol Memoir 7:1–140

Butler AB, Hodos W (2005) Comparative vertebrate neuroanatomy: evolution and adaptation, 2nd edn. Wiley-Interscience, Hoboken

Carlson WD, Rowe T, Ketcham RA, Colbert MW (2003). Geological applications of high-resolution x-ray computed tomography in petrology, meteoritics and palaeontology. In: Mees F, Swennen R, Van Geet M, Jacobs P (eds) Applications of x-ray computed tomography in the geosciences. Geological Society, London, pp 7–22

Chatterjee S, Zheng Z (2002) Cranial anatomy of *Shunosaurus*, a basal sauropod dinosaur from the Middle Jurassic of China. Zool J Linn Soc 136:145–169

Chatterjee S, Zheng Z (2005) Neuroanatomy and dentition of *Camarasaurus lentus*. In: Tidwell V, Carpenter K (eds) Thunder-Lizards: the Sauropodomorph Dinosaurs. Indiana University Press, Bloomington, IN, pp 199–211

Clarke AH (2005) On the vestibular labyrinth of *Brachiosaurus brancai*. J Vestib Res 15:65–71

Colbert EH (1946a) *Sebecus*, representative of a peculiar suborder of fossil Crocodilia from Patagonia. Bull Am Mus Nat Hist 87:217–270

Colbert EH (1946b) The Eustachian tubes in the Crocodilia. Copeia 1946:12–14

Conroy GC, Vannier MW (1984) Noninvasive three-dimensional computer imaging of matrix-filled fossil skulls by high-resolution computed tomography. Science 226:456–458

Domínguez Alonso P, Milner AC, Ketcham RA, Cookson MJ, Rowe TB (2004) The avian nature of the brain and inner ear of *Archaeopteryx*. Nature 430:666–669

Dooling RJ, Lohr B, Dent ML (2000). Hearing in birds and reptiles. In: Dooling RJ, Fay RR, Popper AN (eds) Comparative hearing: birds and reptiles. Springer, New York, pp 308–359

Dubbeldam JL (1998). Birds. In: Nieuwenhuys R, ten Donkelaar HJ, Nicholson C (eds) The central nervous system of vertebrates, Vol. 3. Springer-Verlag, New York City, pp 1525–1636

Edinger T (1942) The pituitary body in giant animals fossil and living: a survey and a suggestion. Q Rev Biol 17:31–45

Edinger T (1975) Paleoneurology 1804–1966. An annotated bibliography. Adv Anat Embryol Cell Biol 49:1–258

Erickson GM, Makovicky PJ, Currie PJ, Norell MA, Yerby SA, Brochu CA (2004) Gigantism and comparative life-history parameters of tyrannosaurid dinosaurs. Nature 430:772–775

Fiorillo AR (1998) Dental microwear patterns of the sauropod dinosaurs *Camarasaurus* and *Diplodocus*: evidence for resource partitioning in the Late Jurassic of North America. Hist Biol 13:1–16

Franzosa JW (2004) Evolution of the brain in Theropoda (Dinosauria). Unpublished PhD dissertation. University of Texas, Austin, Texas, 357 pp.

Franzosa J, Rowe T (2005) Cranial endocast of the Cretaceous theropod dinosaur *Acrocanthosaurus atokensis*. J Vertebr Paleontol 25:859–864

Galton PM (1985) Cranial anatomy of the prosauropod dinosaur *Plateosaurus* from the Knollenmergel (Middle Keuper, Upper Triassic) of Germany. II. All the cranial material and details of soft-part anatomy. Geologica et Palaeontologica 19:119–159

Galton PM (1989) Crania and endocranial casts from ornithopod dinosaurs of the families Dryosauridae and Hypsilophodontidae (Reptilia: Ornithischia). Geologica et Palaeontologica 23:217–239

Galton PM (2001). Endocranial casts of the plated dinosaur *Stegosaurus* (Upper Jurassic, western USA): a complete undistorted cast and the original specimens of Othniel Charles Marsh. In: K Carpenter (ed.) The armored dinosaurs. Indiana University Press, Bloomington, IN, pp 103–129

Gaupp E (1900) Das Chondrocranium von *Lacerta agilis*: ein Beitrag zum Verständnis des Amniotenschädels. Anatomische Hefte 14:435–592 van Gelderen C (1924) Die Morphologie der Sinus durae matris. Erster Teil. Die vergleichende Ontogenie der neurokraniellen Venen der Anamnier und Reptilien. Zeitschrift für Anatomie und Entwicklungsgeschichte 73:541–605

Gilmore CW (1925) A nearly complete articulated skeleton of *Camarasaurus*, a saurischian dinosaur from the Dinosaur National Monument Utah. Mem Carnegie Mus 10:353–367

Gleich O, Manley GA (2000). The hearing organ of birds and Crocodilia. In: Dooling RJ, Fay RR, Popper AN (eds) Comparative hearing: birds and reptiles. Springer, New York, pp 70–138

Gleich O, Fischer FP, Köppl C, Manley GA (2004). Hearing organ evolution and specialization: archosaurs. In: Manley GA, Popper AN, Fay RR (eds) Evolution of the vertebrate auditory system. Springer, New York, pp 224–255

Gleich O, Dooling RJ, Manley GA (2005) Audiogram, body mass, and basilar papilla length: correlations in birds and predictions for extinct archosaurs. Naturwissenschaften 92:595–598

Graf W, Klam F (2006) Le système vestibulaire: anatomie fonctionnelle et comparée, évolution et développement. Compte Rendus Palevol 5:637–655

Gray AA (1908) The labyrinth of animals, including mammals, birds, reptiles and amphibians, Vol. 2. Churchill, London

Henderson DH (1999) Estimating the masses and centers of mass of extinct animals by 3-D mathematical slicing. Paleobiology 25:88–106

Hochstetter F (1906) Beitrage zur Anatomie und Entwickelungsgeschichte des Blutgefassystemes der Krokodile. Reise in Ostafrika. Wissenschaftliche Ergebnisse 4:1–139

Holliday CM, Witmer LM (2007) Archosaur adductor chamber evolution: integration of musculoskeletal and topological criteria in jaw muscle homology. J Morphol 268:457–484

Holliday CM, Ridgely RC, Balanoff AM, Witmer LM (2006) Cephalic vascular anatomy in flamingos (*Phoenicopterus ruber*) based on novel vascular injection and computed tomographic imaging analyses. Anat Rec 288A:1031–1041

Hopson JA (1979). Paleoneurology. In: Gans C (ed.) Biology of the reptilia, Vol 9, Neurology A. Academic Press, New York, pp 39–146

Horner JR (1994) Steak knives, beady eyes, and tiny little arms (a portrait of *T. rex* as a scavenger). In: Rosenberg GD, Wolberg DL (eds) Dino fest. Paleontological Society Special Publication 7, Knoxville, TN, pp 157–164

Hullar TE (2006) Semicircular canal geometry, afferent sensitivity, and animal behavior. Anat Rec 288A:4656–472

Hurlburt GR (1996) Relative brain size in Recent and fossil amniotes: determination and interpretation. PhD Dissertation, University of Toronto, Toronto, Ontario, Canada, 250 pp.

Hurlburt GR, Ridgely RC, Witmer LM (in press). Relative size of brain and cerebrum in *Tyrannosaurus rex*: an analysis using brain-endocast quantitative relationships in extant alligators. In: Parrish JM, Henderson M, Currie PJ, Koppelhus E (eds) Origin, systematics, and paleobiology of the tyrannosauridae. Northern Illinois University Press

Iwaniuk AN, Nelson JE (2002) Can endocranial volume be used as an estimate of brain size in birds. Can J Zool 80:16–23

Janensch W (1935) Die Schädel der Sauropoden *Brachiosaurus, Barosaurus* und *Dicraeosaurus* aus den Tendaguru-Schicten Deutsch-Ostafrikas. Palaeontographica, Supplementum VII 1:147–297

Janensch W (1936) Über Bahnen von Hirnvenen bei Saurischiern und Ornithischiern, sowie einigen anderen fossilen und rezenten Reptilien. Palaeontologische Zeitschrift 18:181–198

Jerison HJ (1973) Evolution of the brain and intelligence. Academic Press, New York

Killian G (1890) Die Ohrmuskeln des Krokodiles nebst vorläufigen Bemerkungen über die Homologie des Musculus stapedius und des Stapes. Jenaische Zeitschrift für Naturwissenschaft 24:632–656

Klump GM (2000) Sound localization in birds. In: Dooling RJ, Fay RR, Popper AN (eds) Comparative hearing: birds and reptiles. Springer, New York, pp 249–307

Knoll F, Galton PM, López-Antoñanzas R (2006) Paleoneurological evidence against a proboscis in the sauropod dinosaur *Diplodocus*. Geobios 39:215–221

Kundrát M (2007) Avian-like attributes of a virtual brain model of the oviraptorid *Conchoraptor gracilis*. Naturwissenschaften 94:499–504

Kurochkin EN, Saveliev SV, Postnov AA, Pervushov EM, Popov EV (2006) On the brain of a primitive bird from the Upper Cretaceous of European Russia. Paleontol J 40:655–667

Larsson HCE (2001) Endocranial anatomy of *Carcharodontosaurus saharicus* (Theropoda: Allosauroidea) and its implications for theropod brain evolution. In: Tanke DH, Carpenter K (eds) Mesozoic vertebrate life. Indiana University Press, Bloomington, IN, pp 19–33

Larsson HCE, Sereno PC, Wilson JA (2000) Forebrain enlargement among nonavian theropod dinosaurs. J Vertebr Paleontol 30:615–618

Macrini TE, Rowe T, Archer M (2006) Description of a cranial endocast from a fossil platypus, *Obdurodon dicksoni* (Monotremata, Ornithorhynchidae), and the relevance of endocranial characters to monotreme monophyly. J Morphol 267:1000–1015

Madsen JH, McIntosh JS Jr, Berman DS (1995) Skull and atlas-axis complex of the Upper Jurassic sauropod *Camarasaurus* Cope (Reptilia: Saurischia). Bull Carnegie Mus Nat Hist 31:1–116

Maisey JG (2005) Braincase of the Upper Devonian shark *Cladoides wildungensis* (Chondrichthyes, Elasmobranchii), with observations on the braincase in early chondrichthyans. Bull Am Mus Nat Hist 288: 1–103

Marino L, Uhen MD, Pyenson ND, Frohlich B (2003) Reconstructing cetacean brain evolution using computed tomography. Anat Rec (Part B) 272B:107–117

Newton ET (1888) On the skull, brain, and auditory organ of a new species of pterosaurian (*Scaphognathus purdoni*), from the Upper Lias near Whitby Yorkshire. Philos Trans R Soc Lond B 179:503–537

Norberg RA (1978) Skull asymmetry, ear Structure and function, and auditory localization in Tengmalm's Owl, *Aegolius funereus* (Linne). Philos Trans R Soc Lond B Biol Sci 282:325–410

Osborn HF (1912) Crania of *Tyrannosaurus* and *Allosaurus*. Mem Am Mus Nat Hist 1:1–30

Osborn HF, Mook CC (1921) *Camarasaurus, Amphicoelias*, and other sauropods of Cope. Mem Am Mus Nat Hist 3:247–387

Ostrom JH, McIntosh JS (1966) Marsh's dinosaurs. The collections from como bluff. Yale University Press, New Haven, CT

Owen R (1850) On the communication between the cavity of the tympanum and the palate in the Crocodilia (gavials, alligators and crocodiles). Philos Trans R Soc Lond 140:521–527

Pickles JO (1988) An introduction to the physiology of hearing, 2nd edition. Academic Press, New York City, NY

Radinsky LB (1968) A new approach to mammalian cranial analysis, illustrated by examples of prosimian primates. J Morphol 124:167–180

Reiner A, Perkel DJ, Bruce L, Butler AB, Csillag A, Kuenzel W, Medina L, Paxinos G, Shimizu T, Striedter GF, Wild M, Ball GF, Durand S, Güntürkün O, Lee DW, Mello CV, Powers A, White SA, Hough G, Kubikova L, Smulders TV, Wada K, Dugas-Ford J, Husband S, Yamamoto K, Siang C, Yu J, Jarvis ED (2004) Revised nomenclature for avian telencephalon and some related brainstem nuclei. J Comp Neurol 473:377–414

Rogers SW (1998) Exploring dinosaur neuropaleobiology: viewpoint computed tomography scanning and analysis of an *Allosaurus fragilis* endocast. Neuron 21:673–679

Rogers SW (1999) *Allosaurus*, crocodiles, and birds: evolutionary clues from spiral computed tomography of an endocast. Anat Rec 257:162–173

Rogers SW (2005) Reconstructing the behaviors of extinct species: an excursion into comparative paleoneurology. Am J Med Genet 134A:349–356

Sampson SD, Witmer LM (2007) Craniofacial anatomy of *Majungasaurus crenatissimus* (Theropoda: Abelisauridae) from the Late Cretaceous of Madagascar. Society of Vertebrate Paleontology Memoir 8. J Vertebr Paleontol 27(Suppl. 2):32–102

Sedlmayr JC (2002) Anatomy, evolution, and functional significance of cephalic vasculature in Archosauria. Ph. D. dissertation, Ohio University, 398 pp

Shiino K (1914) Studien zur Kenntnis des Wirbeltierkopfes. I. Das Chondrocranium von *Crocodilus* mit Berücksichtigung der Gehirnnerven und der Kopfgefässe. Anatomische Hefte 50:257–381

Spoor F (2003) The semicircular canal system and locomotor behaviour, with special reference to hominin evolution. In: Franzen JL, Köhler M, Moyà-Solà S (eds), Walking upright. Courier Forschungsinstitut Senkenberg 243, Stuttgart, pp 93–104

Spoor F, Zonneveld F (1998) Comparative review of the human bony labyrinth. Yearb Phys Anthropol 41:211–251

Spoor F, Garland T Jr, Krovitz G, Ryan TM, Silcox MT, Walker A (2007) The primate semicircular canal system and locomotion. Proc Natl Acad Sci 104:10808–10812

Stensiö EA (1963) The brain and cranial nerves in fossil lower craniate vertebrates. Skrifter utgitt av Det Norske Videnskaps-Akademi i Oslo. I. Matematisk-naturvidenskapelig klasse Ny; serie no 13, pp 1–120

Stevens KA, Parrish JM (2005a) Digital reconstructions of sauropod dinosaurs and implications for feeding. In: Curry Rogers K, Wilson JA (eds) The sauropods: evolution and paleobiology. University of California Press, Berkeley, CA, pp 178–200

Stevens KA, Parrish JM (2005b) Neck posture, dentition, and feeding strategies in Jurassic sauropod dinosaurs. In: Tidwell V, Carpenter K (eds) Thunder-lizards: the sauropodomorph dinosaurs. Indiana University Press, Bloomington, IN, pp 213–232

Stingelin W (1957) Vergleichend morphologische Untersuchungen am Verderhirn der Vögel auf cytologischer und cytoarchitektonischer Grundlage. Verlag Helbing & Lichtenhahn, Basel

Tanturri V (1933) Zur Anatomie und Physiologie des Labyrinthes der Vögel. Monatsschrift für Ohrenheilkunde und Laryngo-Rhinologie 67:1–27

Turkewitsch BG (1934) Zur Anatomie des Gehörorgans der Vögel (Canales semicirculares). Zeitschrift für Anatomie und Entwicklungsgeschischte 103:551–608

Turner CH (1891) Morphology of the avian brain. J Comp Neurol 1:39–92, 265–286

Weishampel DB, Barrett PM, Coria RA, Loeuff J. Le, Xu X, Zhao X, Sahni A, Gomani EMP, Noto CR (2004) Dinosaur distribution. In: Weishampel DB, Dodson P, Osmólska H (eds), The dinosauria, 2nd edn. University of California Press, Berkeley, CA, pp 517–606

Wever EG (1978) The reptile ear. Its structure and function. Princeton University Press, Princeton, NJ

Wharton DS (2002) The evolution of the avian brain. Ph. D. dissertation, University of Bristol, UK

White TR (1958) The braincase of *Camarasaurus lentus* (Marsh). J Paleontol 32:477–494

Witmer LM (1990) The craniofacial air sac system of Mesozoic birds (Aves). Zool J Linn Soc 100:327–378

Witmer LM (1995). The Extant Phylogenetic Bracket and the importance of reconstructing soft tissues in fossils. In: Thomason JJ (ed.) Functional morphology in vertebrate paleontology. Cambridge University Press, New York, pp 19–33

Witmer LM (1997) Craniofacial air sinus systems. In: Currie PJ, Padian K (eds) Encyclopedia of dinosaurs. Academic Press, New York, pp 151–159

Witmer LM, Ridgely RC (in press) The Cleveland tyrannosaur skull (*Nanotyrannus* or *Tyrannosaurus*): new findings based on CT scanning, with special reference to the braincase. Kirtlandia

Witmer LM, Chatterjee S, Franzosa J, Rowe T (2003) Neuroanatomy of flying reptiles and implications for flight, posture and behavior. Nature 425:950–953

7 Evolutionary Morphology of the Autonomic Cardiac Nervous System in Non-human Primates and Humans

Tomokazu Kawashima and Hiroshi Sasaki

7.1 Introduction

The autonomic cardiac nervous system (ACNS) is responsible for important cardiac functions, such as heart pulse, heart contraction, heart pressure, and cardiac pain delivery. Therefore, many studies of the ACNS from viewpoints of the clinical anatomy and functional analyses using experimental animals have been conducted.

In the past decade, imaging devices, such as high-resolution CT and 3D-reconstruction equipment, have revealed numerous significant spatial morphological details without damaging the specimen, even in the field of anatomical research. However, both imaging analysis and anatomical dissection have inherent problems in revealing the very complicated peripheral autonomic nervous system owing to difficulties in distinguishing it from connective tissue. Irrespective of this difficulty, its detailed morphologies need to be investigated in every field of clinical medicine to get a better understanding of its precise morphology and topographic relationships.

On the other hand, various studies on the ACNS have been performed on the basis of morphological, physiological, and pharmacological approaches. Experimental stdies used model animals such as rats, mice, or dogs; nevertheless, the ACNS morphologies show considerable interspecific differences. The main trunk of the cervical sympathetic nerve and the vagus nerve, e.g., are fused as one trunk, the vagosympathetic trunk, in the dog (Nonidez 1939; Mizeres 1955, 1958). Recently, most of the ACNS research has shifted into the microscopic histochemical and molecular embryological levels (Pauza et al. 2000; Verberne et al. 1998, 2000a, 2000b). This means that the ACNS research focuses narrowly on regional details, despite the fact that there are several morphological aspects that still require clarification at the macroscopic level. In addition, the research most practically demanded for clinicians would be the macroscopic anatomy. On the other hand, anatomical reexaminations of the autonomic nervous system in the abdominal and pelvic regions have been performed for function-preserving operations aimed at improving patient's quality of life. In cardiac surgery, function-preserving operations have never been conducted because of their difficulty. Since such a kind of surgery is anticipated in the future, detailed anatomical investigation of the human ACNS will be required.

We have examined the ACNS in some species, including humans. Regarding humans, not only the general morphology and its normal (typical) variation (Kawashima 2005) but also topographical changes of the ACNS deriving from branchial artery anomaly (Kawashima and Sasaki 2005, 2007), have been described. To understand completely the morphological principles of the ACNS, it is essential to examine the ACNS also in several animals and compare it with humans on a phylogenetic and evolutionary basis. Therefore, we have studied the comparative anatomy of the ACNS in Asian Cercopithecidae, macaque monkeys (Kawashima et al. 2005) and in African Cercopithecidae (Kawashima et al. 2007). These results show significant morphological differences between humans and non-human primates, which are phylogenetically much closer to humans than experimental animals such as mouse, rat, and dog.

Department of Anatomy, School of Medicine, Tokyo Women's Medical University, Tokyo 162-8666, Japan

While various molecular DNA studies on the evolution of primates have revealed their phylogenic relationships, macroscopic evolutionary studies of the ACNS have discovered many unclear points. Therefore, we have attempted to examine evolutionary transformations of the ACNS in primates, including some species not recorded so far. The results have implications not only for evolutionary but also for paleontological studies. In this chapter, we will describe the comparative anatomy of the ACNS in primates, point out interspecific differences, and try to relate these to their morphological and functional significance.

7.2 Description

7.2.1 Definition of the Autonomic Cardiac Nervous System

A lot of technical terms on the anatomy of the ACNS have been introduced arbitrarily without clear definitions. Moreover, since these technical terms apply mainly to generalized anatomical structures, they are not sufficient for describing interspecific/inter-individual differences. As a result, authors have separately coined new terms, leading to great difficulties in comparing results among previous literature (Nonidez 1939; Mitchell 1953; Hoffman 1957; Pick 1970; Fukuyama 1982).

We clearly classified the nerves innervating the heart into the sympathetic cardiac nerves and the parasympathetic vagal cardiac branches according to Nomina Anatomica (Fig. 7.1A).

Within the cardiac nerves connecting to the cardiac plexus or distributing to the heart itself, each cardiac nerve is named according to its site of origin as follows:

superior cardiac nerves, cardiac nerves originating from the superior cervical ganglion or the sympathetic trunk between the superior and the middle cervical ganglia

middle cardiac nerves, cardiac nerves originating from the (accessory) middle cervical, vertebral ganglia or the sympathetic trunk between the middle and inferior cervical ganglia (or the cervicothoracic ganglion), including the *ansa subclavia*

inferior cardiac nerves, cardiac nerves originating from the inferior cervical or cervicothoracic (stellate) ganglia

thoracic cardiac nerves, cardiac nerves originating from the thoracic ganglia or the thoracic sympathetic trunk below the inferior cervical or the cervicothoracic ganglia.

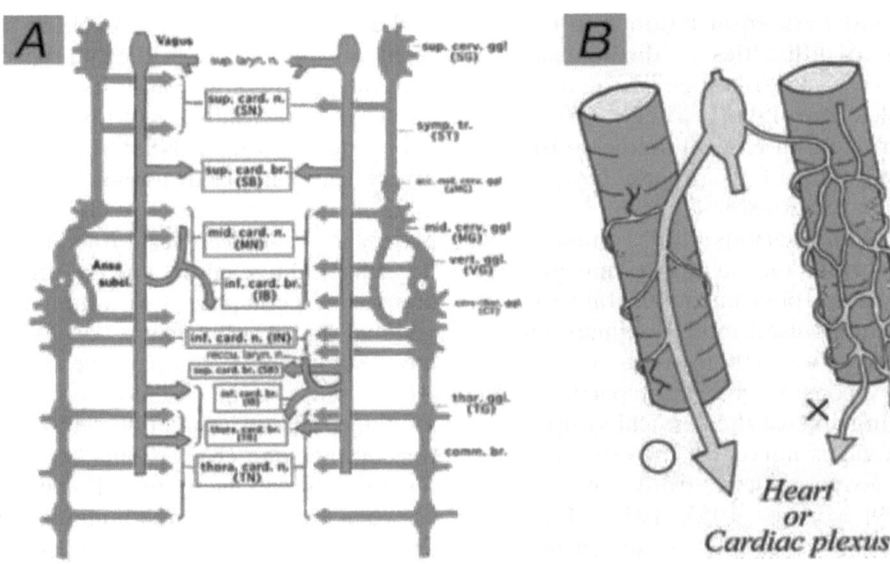

FIG. 7.1. Diagrams showing our definition of the autonomic cardiac nervous system. (**A**) Each cardiac nerve/branch is named according to its origin. The sympathetic cardiac nerves and the parasympathetic vagal cardiac branches are colored in *orange* and *green*, respectively. (**B**) The cardiac nerves/branches are regarded as nerves with direct connections via the cardiac plexus, except for slender thin nerves/branches like the sensory nerves around great vessels (see Color Plates, Fig. 7.1).

Each cardiac branch is named according to its site of origin from a developmental viewpoint, as follows:

superior cardiac branch, cardiac branch arising from the vagus nerve in the upper (proximal) portion of the recurrent laryngeal nerve

inferior cardiac branch, cardiac branch originating from the recurrent laryngeal nerve

thoracic cardiac branch, cardiac branch originating from the vagus nerve of the lower (distal) portion of the recurrent laryngeal nerve

In addition, we restricted the use of these cardiac nerves/branches to those easily distinguishable from the surrounding connective tissues excepting the slender thin nerves/branches like sensory nerves around great vessels, finally connecting to the cardiac plexus or distributing to the heart directly (Fig. 7.1B).

7.2.2 Extrinsic Autonomic Cardiac Nervous System

7.2.2.1 Old World Monkeys (*Cercopithecidae*)

Many papers on the ACNS of this taxonomic group have been published (Sonntag 1922; Riegele 1925–1926; Zuckerman 1938; Botár and Becker 1939; Sheehan and Pick 1943; Botár et al. 1950; Pick 1970; Kawashima and Sato 2000; Kawashima et al. 2001, 2005, 2007). The respective numbers of vertebrae show some variation. The cervical vertebrae, of course, are consistently seven as in the majority of other mammals, whereas the number of thoracic vertebrae varies between twelve and thirteen.

The ACNS in Old World monkeys shows morphological variation (van den Broek 1908). From an embryological viewpoint, the superior, middle, and inferior cervical and 12 or 13 thoracic ganglia are expected to be present. The superior cervical and middle cervical ganglia, the cervicothoracic ganglion composed of the inferior cervical, first and second thoracic ganglia, and the third to 12th (13th) thoracic ganglia are usually observed (Figs 7.2,7.3).

Until now, the sympathetic ganglia have been normally identified by examination of the communicating branches with the spinal nerves (Harman 1900; Axford, 1927–28; Pick and Sheehan 1946; Jamieson et al. 1952; Mitchell 1953; Becker and Grunt 1957; Hoffman 1957; Wrete 1959). The superior cervical, the cervicothoracic, and each of the thoracic ganglia communicate with C1-(2)3, C8-T2(3), and each of the segmental thoracic nerves, respectively. On the other hand, the middle cervical ganglion does not communicate with any spinal nerves (van den Broek 1908; Pick 1970; Kawashima et al. 2005). A middle cervical ganglion in Old World Monkeys has not yet been identified. It suggests that there are vacant cervical segments without communicating branches from the sympathetic cervical ganglia. In this taxonomic group, the vacant cervical segments were covered by the vertebral nerve which mostly originated from the cervicothoracic ganglion, ran along the vertebral artery, and connected not only to the middle cervical segments but also to superior cervical segments in an overlapping area with the communicating branches from the superior cervical ganglion. On the basis of these findings, Wrete (1959) regarded the vertebral nerve as the deep sympathetic trunk.

The main sympathetic cardiac nerves are the middle cardiac nerve originating from the middle cervical ganglion and the inferior cardiac nerve originating from the cervicothoracic ganglion.

In addition, about 50% of the cardiac nerves originating from the sympathetic trunk between the superior and middle cervical ganglia and 25% of the thoracic cardiac nerves also participate in the cardiac innervation (Kawashima et al. 2005, 2007). In our observation, the superior cardiac nerve originating from the superior cervical ganglion, which has never been documented for Cercopithecidae in the available literature, was observed in only one side of one Anubis baboon (*Papio anubis*) as shown in Fig. 7.3B. This finding suggests that the superior cardiac nerve originating from the superior cervical ganglion is normally absent in Old World monkeys. On the other hand, all parasympathetic vagal cardiac branches (the superior, inferior, and thoracic cardiac branches of the vagus nerve) are consistently observed regardless of the species.

During the morphogenesis of the autonomic cardiac nerves, the vagal cardiac branches extend to the heart along the branchial arches at first, later followed by the partial disappearance of the branchial arches, arteries, and related organs. Thereafter, the sympathetic cardiac nerves

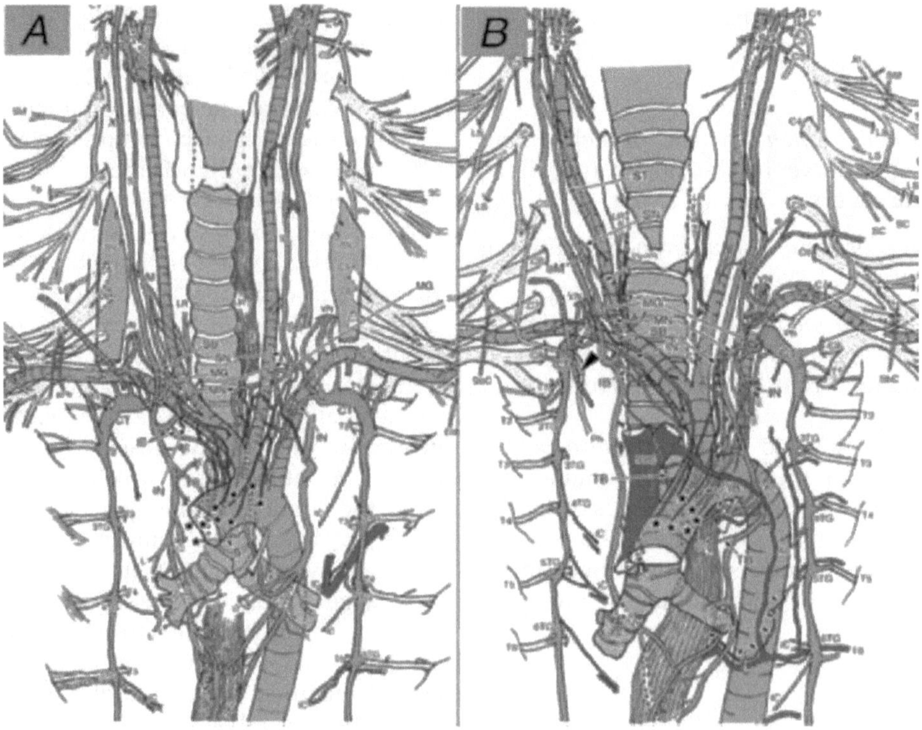

FIG. 7.2. Diagram showing the autonomic cardiac nervous system and its surrounding structure in Asian Old World monkeys *(Cercopithecidae)*. (**A**) rhesus monkey *(Macaca mulatta)*. (**B**) stump-tailed monkey (*Macaca arctoides*), ventral view. The black and white stars show the cardiac nerves/branches entering from the arterial and venous parts of the pericardium reflection, respectively (After Kawashima et al. 2005). aM, accessory middle ganglion; CT, cervicothoracic (stellate) ganglion; IB, inferior (vagal) cardiac branch; IN, inferior cervical cardiac nerve; L, nerve branches to lung; LR, recurrent laryngeal nerve of vagus nerve; MG, middle cervical ganglion; MN, middle cardiac nerve; Ph, phrenic nerve; SVC, superior vena cava; SB, superior (vagal) cardiac branch; SbC, nerve to subclavian muscle; SC, supraclavicular nerve; SG, superior cervical ganglion; SM, nerve branch to sternocleidomastoid muscle; SN, superior cardiac nerve; SS, suprascapular nerve; ST, sympathetic trunk; TB, thoracic (vagal) cardiac branch; Tp, nerve branch to trapezius muscle; VN, vertebral nerve; *X*, vagus nerve; *XI*, accessory nerve; 2-6TG, second to sixth thoracic ganglia (see Color Plates, Fig. 7.2).

secondarily extend to the heart, forming the cardiac plexus when the morphogenesis of the derivatives of the branchial arteries is nearly complete (Sharner 1930; Gomez 1958; Kuratani et al. 1991; Roberts 1991). These embryological results support our anatomical findings of the parasympathetic vagal cardiac branch dominance in the Old World monkeys.

7.2.2.2 Apes

In a long anatomical history, the data on the ACNS in apes are insufficient and strongly require further study (van den Broek 1908; Sonntag 1924; Botar 1932; Kolesnikov 1935). Some authors state that in the apes the superior cardiac nerve arises from the superior cervical ganglion (gorilla by Botar 1932); orangutan by (Kolesnikov 1935), whereas other authors have opposed to these views (white-handed gibbon and orangutan by van den Broek 1908; orangutan by Sonntag 1924; chimpanzee by Kolesnikov 1935). Thus, it is necessary to re-examine the ACNS in apes.

In our experiences, the superior cervical and middle cervical ganglia, the vertebral ganglion, the cervicothoracic ganglion composed of the inferior cervical and first thoracic ganglia, and the third to 12th (13th) thoracic ganglia are usually present.

The superior cervical, the cervicothoracic, and each of the thoracic ganglia communicate with C1-(2)3, C8-T2(3), and each segmental and a lower thoracic nerves, respectively. In contrast, the presence of the communicating branches from the middle cervical and vertebral ganglia to the cervical nerve is not consistent. The vertebral nerve tends to originate from the vertebral and

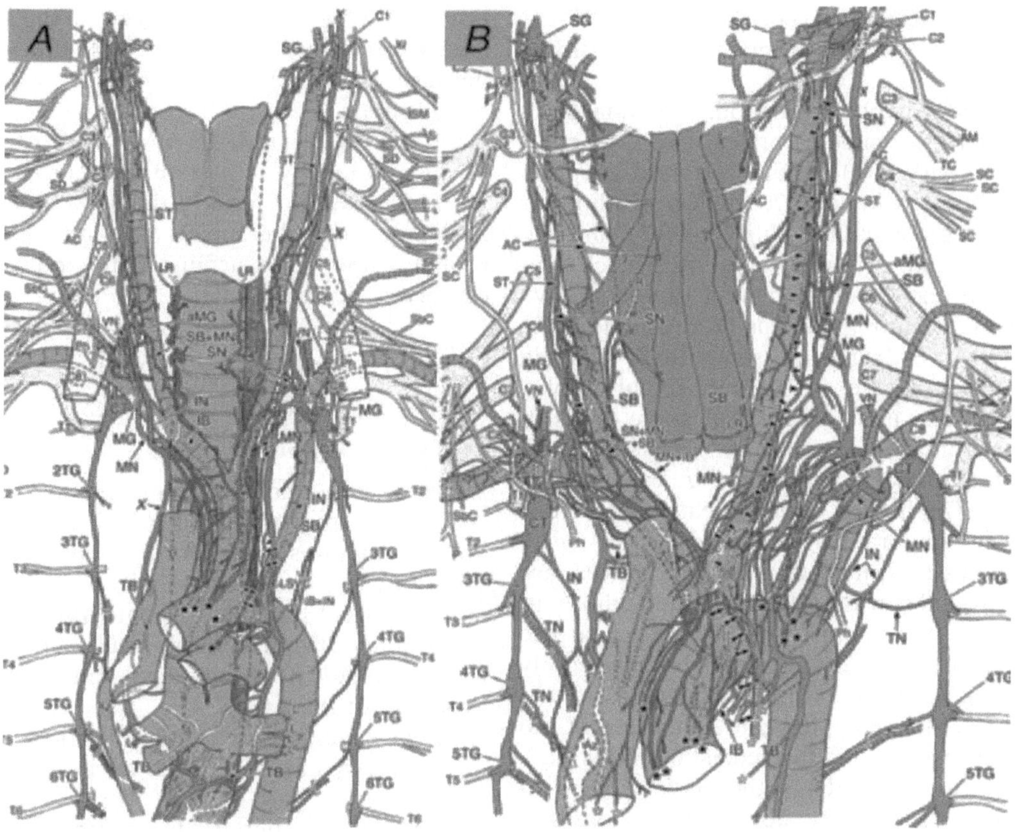

FIG. 7.3. Diagrams showing the autonomic cardiac nervous systems in African Old World monkeys *(Cercopithecidae)*. (**A**) Savanna monkey *(Chlorocebus aethiops)*. (**B**) Anubis baboon *(Papio anubis)*. Viewed from the ventral aspect. The black and white stars show the cardiac nerves/branches entering from the arterial and venous parts of the pericardium reflection, respectively (After Kawashima et al. 2007). AC, cervical ansa; AM, greater auricular nerve; aMG, accessory middle ganglion; CT, cervicothoracic (stellate) ganglion; IB, inferior (vagal) cardiac branch; IN, inferior cervical cardiac nerve; L, nerve branches to lung; LR, recurrent laryngeal nerve of vagus nerve; MG, middle cervical ganglion; MN, middle cardiac nerve; P, pectoral nerve; Ph, phrenic nerve; SB, superior (vagal) cardiac branch; SbC, nerve to subclavian muscle; SC, supraclavicular nerve; SD, dorsal scapular nerve; SG, superior cervical ganglion; SM, nerve branch to sternocleidomastoid muscle; SN, superior cardiac nerve; ST, sympathetic trunk; TB, thoracic (vagal) cardiac branch; TC, transverse cervical nerve; TN, thoracic cardiac nerve; VN, vertebral nerve; *X*, vagus nerve; *XI*, accessory nerve; *XII*, hypogastric nerve; 2-6TG, second to sixth thoracic ganglia (see Color Plates, Fig. 7.3).

cervicothoracic ganglia, but not from the middle cervical ganglion. In addition, the communicating segments of the vertebral nerve in apes are weaker than those in Old World monkey. They normally communicate with the middle and lower cervical nerves.

As in the Old World monkeys, the main sympathetic cardiac nerves in apes are forming the middle cardiac nerve, originating from the middle cervical ganglion, and the inferior cardiac nerve, originating from the cervicothoracic ganglion. Nevertheless, a superior cardiac nerve, originating from the superior cervical ganglion, is often observed in apes. On the other hand, all parasympathetic vagal cardiac branches are consistently present as in the Old World monkeys.

Further studies in apes are required to supplement our fragmental observations and few references in the literatures.

7.2.2.3 Humans

The regular number of thoracic vertebrae in humans is twelve, although we sometimes encounter 13 thoracic vertebrae and ribs (2.0–6.1%, Hirata 2000). Detailed anatomical investigations of the human ACNS have been conducted (Mitchell 1953; Pick 1970; Fukuyama 1982; Kawashima 2005).

The superior cervical ganglion, communicating with C1-3(4), the middle cervical ganglion, communicating with C(3)4-6, the vertebral ganglion communicating with C(6)7, the cervicothoracic ganglion communicating with C7-T1(2), and the thoracic ganglia, communicating with each segmental and one lower segmental thoracic nerve are consistent among humans (Kawashima 2005; Fig. 7.4). Accordingly, all sympathetic ganglia always communicate with some spinal nerves. In addition, the human vertebral nerve tends to originate from the vertebral and cervicothoracic ganglia, but not from the middle cervical ganglion, as in apes. The highest communicating branch of the human vertebral nerve is C5 at best.

The human sympathetic cardiac nerves are completely and consistently present in addition to the presence of all parasympathetic vagal cardiac branches. In these cardiac nerves/branches, only the thoracic cardiac nerve in the posterior mediastinum follows a complex course because of the long distance to the middle mediastinum (Fig. 7.5). According to our findings, the actual course of the right thoracic cardiac nerve differs from that provided by previous descriptions (Kuntz and Morehouse 1930; Saccomanno 1943).

7.2.3 Intrinsic Autonomic Cardiac Nervous System

The cardiac inlet/outlet entrances were limited due to the pericardium, arterial and venous parts of the *hilum cordis*, vacancy of the pericardium on the dorsal aspect, and the *hilum cordis* itself (Pauza et al. 1997). The arterial part (ascending aorta and pulmonary trunk) and a part of the venous part (superior vena cava) tended to enter and leave a relatively thick cardiac nerve/branch arising from the main cranial trunks, whereas most venous parts (pulmonary veins and inferior vena cava) tended to enter and leave the relatively thin cardiac nerves/branches from the caudal aspect (Fig. 7.6).

The arterial group deriving form the superior, middle, and inferior cardiac nerves, innervating the heart, form a great neural plexus and distribute to the heart by simply following the descent of the great arteries. In contrast, the venous group derives from the weak thoracic cardiac nerve and branch. The thoracic cardiac nerve in the posterior mediastinum has to follow a complex course because of the long distance to the middle mediastinum.

On the other hand, the *hilum cordis* tended to be distributed by a part of the cardiac plexus. Although the total mass of the nerve fibers of the ACNS seems to be subject to some variation in primates, the basic arrangement and structural characteristics are similar.

7.2.4 Evolutionary Morphology of the Autonomic Cardiac Nervous System

As described above, the morphologies of the ACNS seem to be rapidly generated and differentiated structures in later developmental stage. Therefore, by discussing their phylogeny on a morphological basis and by integrating the findings obtained from both molecular and morphological studies, we might shed some light on the evolution of the ACNS.

In particular, the morphology and composition of the cervicothoracic ganglion in primates are interesting in this respect (Fig. 7.7). Results suggest an evolutionary transition from a relatively large cervicothoracic ganglion, composed of the inferior cervical, the first thoracic, and the second thoracic ganglia, to a single inferior cervical ganglion.

On the other hand, all parasympathetic vagal cardiac branches are seen consistently whereas the sympathetic cardiac nerves extended their range of origin cranially and caudally in correspondence with their developmental features.

7.2.5 Topographical Changes of the Human Autonomic Cardiac Nervous System and Their Anatomical Significance

In order to comprehend the general architecture of the ACNS, it is necessary to pay attention not only to the topographical changes of the ACNS but also to its relation with accompanying great vessels. The arterial system and the peripheral autonomic nervous system have been studied separately because of variations in each system and the difficulty in identifying causal relationships between various morphogenetic factors interacting in the

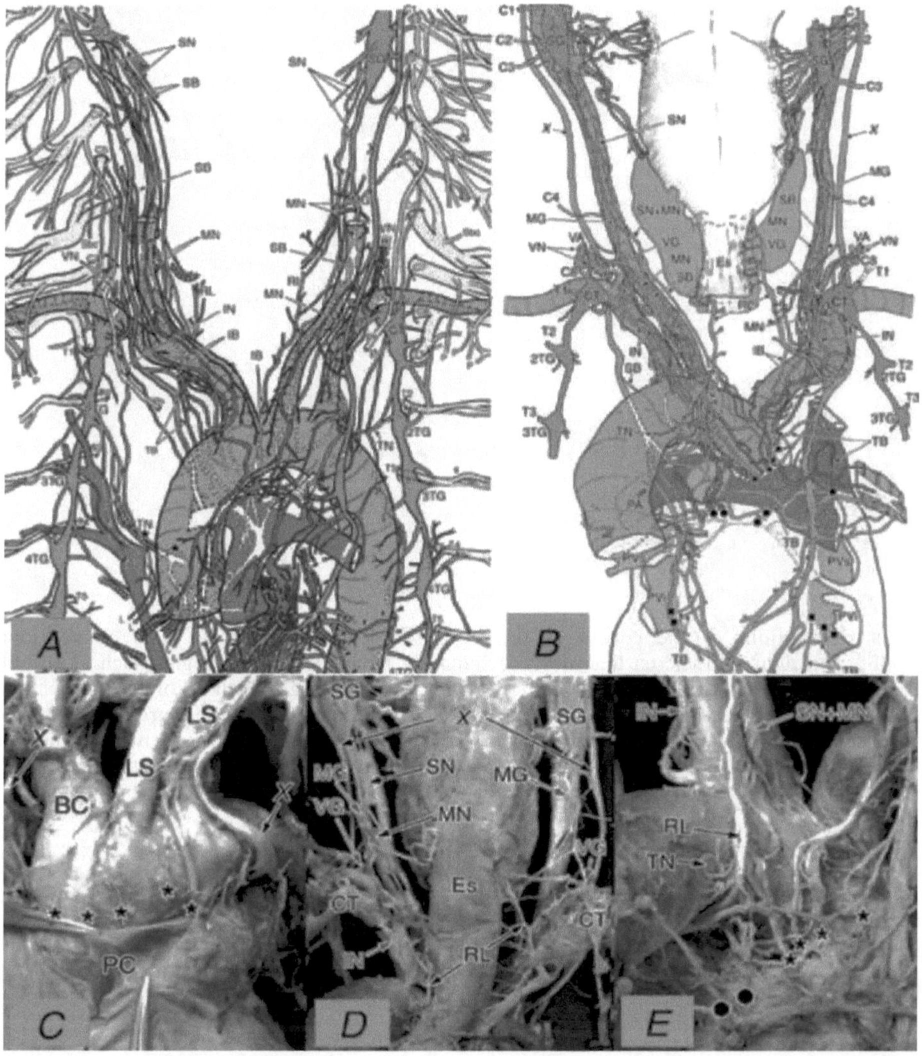

Fig. 7.4. Diagrams and photographs showing the autonomic cardiac nervous systems (ACNS) in humans. (**A**) Ventral view. (**B**) Dorsal view. The sympathetic cardiac nerves, vagal cardiac branches, and the mixture nerves (cardiac plexus) are colored orange, green, and purple, respectively. (**C**) Ventral view. The reflection of the pericardium in the arterial part of the *hilum cordis* and its nerve entrances are clearly observed with incision of a part of the parietal pericardium and its retraction by forceps. (**D**) Dorsal view. The ACNS accompanying the great arteries. (**E**) The complex cardiac plexus. An enlargement of Fig. 7.1B. As seen in this case, all cardiac nerves and branches are normally observed along the great vessels. Black stars (★), squares (■), and circles (•) indicate the inlet/outlet of the cardiac nerves/branches from the arterial, venous parts of, and the *hilum cordis* (After Kawashima 2005; Kawashima and Sasaki 2006). CT, cervicothoracic (stellate) ganglion; Es, esophagus; IB, inferior (vagal) cardiac branch; IN, inferior cervical cardiac nerve; LC, left common carotid artery; LS, left subclavian artery; MG, middle cervical ganglion; MN, middle cardiac nerve; PA, pulmonary artery; PC, pericardial cavity; PVi, inferior pulmonary vein; PVs, superior pulmonary vein; RL, recurrent laryngeal nerve; SB, superior (vagal) cardiac branch; SbC, branch to the subclavian muscle; SG, superior cervical ganglion; SN, superior cardiac nerve; SVC, superior vena cava; TB, thoracic (vagal) cardiac branch; TG, thoracic ganglia; TN, thoracic cardiac nerve; VA, vertebral artery; VG, vertebral ganglion; VN, vertebral nerve; *X*, vagus nerve (see Color Plates, Fig. 7.4).

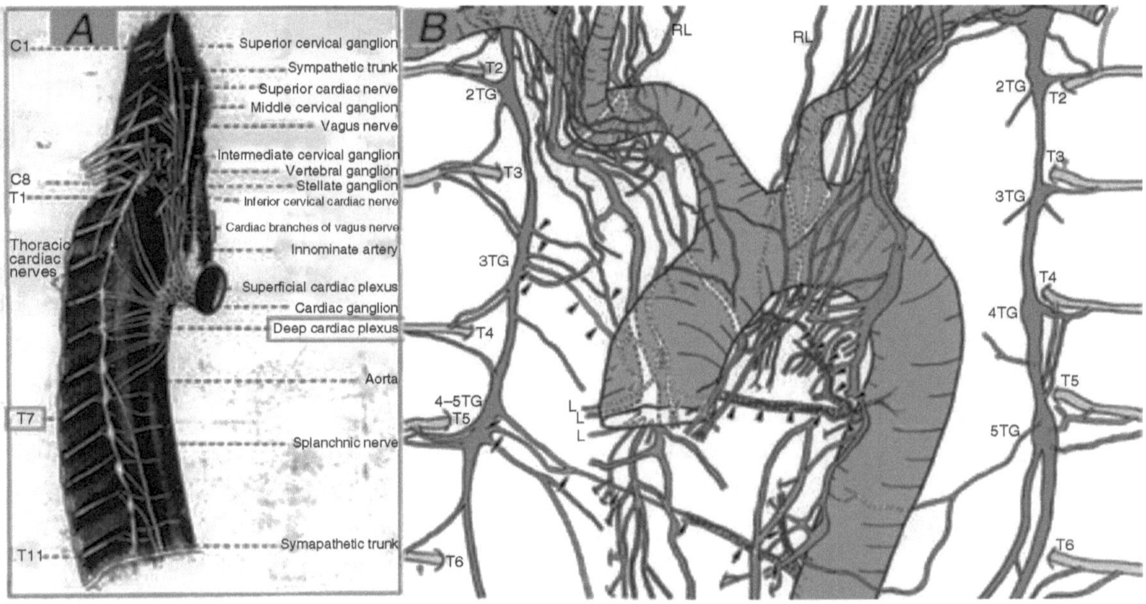

FIG. 7.5. Different descriptions of the course in the right thoracic cardiac nerve. (**A**) conventional description (Saccomanno 1943). The right thoracic cardiac nerve originating from the lower ganglia, such as the fifth to seventh thoracic ganglia, ascends to reach the heart at a position higher than the origin of the thoracic cardiac nerve, regardless of the course of the intercostal vessels. (**B**) Our new description (recurrent right thoracic cardiac nerve: Kawashima 2005; Kawashima and Sasaki 2005). It descends obliquely along the intercostal vessels. After arriving at the thoracic aorta, it turns there and ascends along the aorta, finally distributing to the heart via the left venous part of cardiac hilum or connecting to the cardiac plexus as shown by the arrow heads (see Color Plates, Fig. 7.5).

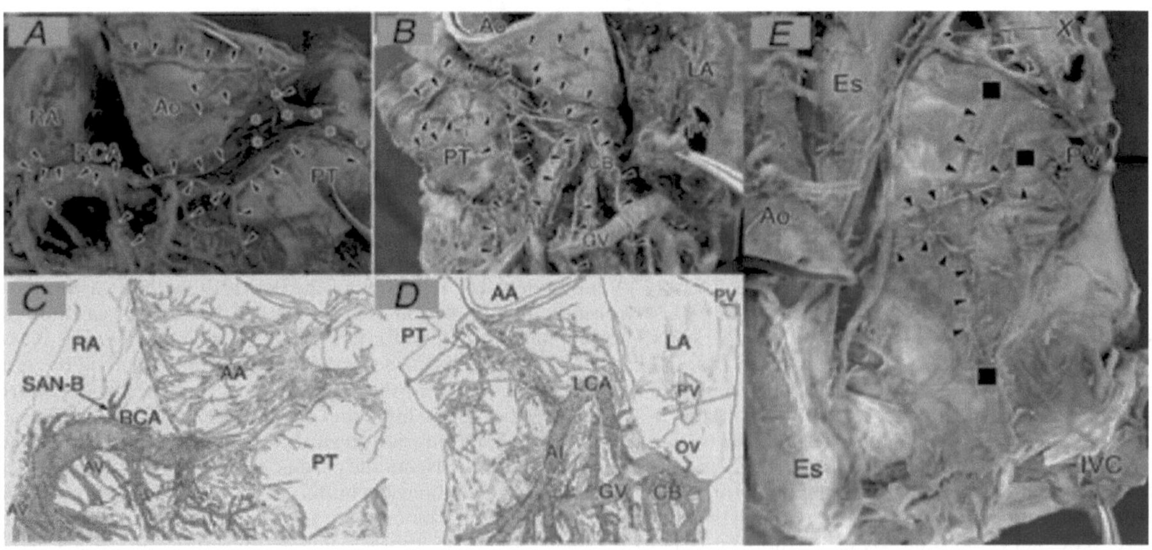

FIG. 7.6. Photographs and diagrams showing the intrinsic peripheral cardiac distributions. (**A, C**) Arterial group consisting of relatively thick cardiac nerves/branches. Right sides. (**B, D**) Left side. The bundle (*) of the cardiac plexus courses between the aorta and the pulmonary trunk, runs along the great vessels and both coronary arteries, and then distributes to the atria, ventricles, and coronary arteries. (**E**) Venous group consisting of very thin branches, dorsal view. Arrow heads and squares (■) show the course and entrance to the pericardium, respectively (After Kawashima 2005). AI, anterior interventricular branch of left coronary artery; CB, circumflex branch of left coronary artery; Es, esophagus; GV, great cardiac vein; IVC, inferior vena cava; LA, left atrium; LCA, left coronary artery; PV, pulmonary vein; RA, right atrium; RCA, right coronary artery; X, vagus nerve (see Color Plates, Fig. 7.6).

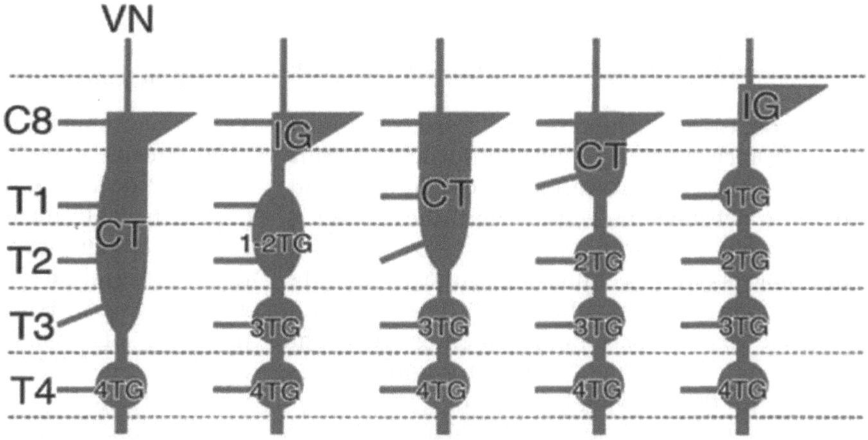

Species	sides	CT(IG+ 1-3TG)	IG+ 1-2TG	CT(IG+ 1-2TG)	CT(IG+ 1TG)	IG+1TG-
Japanese monkey (*Macaca fuscata*)	20	3	2	14	1	0
Rhesus monkey (*Macaca mulatta*)	4	0	0	4	0	0
Crab-eating monkey (*Macaca fascicularis*)	10	0	0	6	4	0
Stump-tailed monkey (*Macaca arctoides*)	2	0	0	1	1	0
Pig-tailed monkey (*M. nemestrina nemestrina*)	4	0	0	2	2	0
Pig-tailed monkey (*M. nemestrina leonina*)	2	0	0	2	0	0
Abyssinian colobus (*Colobus guereza*)	2	0	1	1	0	0
Angola pied colobus (*Colobus angolensis ruwenzorii*)	2	0	0	1	1	0
Savanna monkey (*Chlorocebus aethiops*)	4	0	0	2	2	0
Lesser white-nosed Guenon (*Cercopithecus petaurista*)	4	0	0	3	1	0
Sacred baboon (*Papio hamadryas*)	2	0	0	1	1	0
Anubis baboon (*Papio anubis*)	6	0	0	6	0	0
Mandrill (*Mandrillus sphinx*)	2	0	0	2	0	0
Chimpanzee (*Pan troglodytes*)	2	0	0	0	2	0
Human (*Homo sapiens*)	36	0	0	1	30	5

Fig. 7.7. Evolutionary comparison of the cervicothoracic ganglion among primates. CT, cervicothoracic ganglion; IG, inferior cervical ganglion; VN, vertebral nerve; 1TG-4TG, first to fourth thoracic ganglia.

two systems. However, Kasai (1962) emphasized that it is very important to analyze relationship between the two systems, especially in cases with arterial anomalies. We approached this problem by examining ACNS variation while using a typical great arterial branching pattern from the aortic arch as a control.

7.2.5.1 Retroesophageal Right Subclavian Artery (RERSA)

The presence of a retroesophageal right subclavian artery (RERSA) has been reported by many authors because of its relatively high frequency (0.1–1.6%), compared with other rare human branchial arterial variations and its eventual occurrence has been confirmed recently (Nakatani et al. 1996; Kurt et al. 1997; Shimada et al. 1997; Jakubowicz et al. 2002; Loukas et al. 2004).

In most cases, the first, second, third, and fourth (last) branches arising from the aortic arch were the right common carotid, the left common carotid, the left subclavian arteries, and the RERSA, respectively. According to recent experimental teratology as well as embryological textbooks, the retroesophageal right subclavian artery is formed from the right seventh intersegmental artery and a distal part of the right dorsal aorta after interruption of the proximal part, i.e., the third to seventh segments, of the right dorsal aorta (Wilson and Warkany 1949, 1950; Kitchell et al. 1957; Okishima et al. 1987; Sumida 1988; Tasaka et al. 1991). Therefore, a corresponding and consistent lack of the right recurrent laryngeal nerve of the vagus nerve is easily understandable based on branchial artery development. However, there are quite a few reports about the spatial changes of the ACNS in individuals with a RERSA (Kumaki 1980; Chiba et al. 1981; Horiguchi et al. 1982; Suzuki et al. 1990; Tanaka 2000). The topological changes of the human ACNS with a RERSA were examined in detail by comparing it with the typical ACNS (Kawashima and Sasaki 2005). The following new results were obtained in addition to the common finding of a lack of the right recurrent laryngeal nerve.

According to previous reports on a RERSA, a thick right thoracic cardiac nerve is commonly observed instead of a right superior cardiac nerve. However, all the cardiac nerves were recognized by us in individuals with RERSA. Thus, we strongly oppose the previous idea that the origin of the right cardiac nerves from the sympathetic trunk and ganglia might have been shifted caudally in individuals with RERSA. In addition, the right thoracic cardiac nerves, which have never been observed to accompany the right subclavian artery, ran along the proximal part of the RERSA. Accordingly, these findings indicate that the proximal part of the RERSA may originate caudal to the proximal part of the normal subclavian artery based on the cardiac nervous morphology of the following the great arteries as well as the morphogenesis of the RERSA.

The topological changes of the autonomic cardiac nervous system in two cases of the RERSA also reflected spatial changes of the great arteries.

7.2.5.2 Anomalous Left Vertebral Artery

We present a further anomaly of the human branchial arteries, namely an anomalous left vertebral artery. Proceeding from proximal to distal, the great arterial branches of the aortic arch were the brachiocephalic trunk, the left common carotid, the left vertebral (VA), and the left subclavian arteries (Fig. 7.8B). The right VA branched from the right subclavian artery and entered into the usual sixth cervical transverse vertebral foramen, whereas the left VA entered into the fourth. Therefore, the left VA showed an anomalous origin and entrance. All cardiac nerves/branches were bilaterally observed.

Remarkably, significant differences in the sympathetic ganglion and cardiac nerves were found on the left side. The cranial portion of the cervicothoracic (stellate) ganglion, composed of the inferior cervical and first thoracic ganglia, was outstanding and constricted, as shown by the large arrow in Fig. 7.8B. In addition, thick middle and inferior cardiac nerves arising from the middle cervical ganglion and from the cranial portion of the cervicothoracic ganglion coursed along the anomalous left VA. So did in addition to the superior cardiac nerve, arising from the superior cervical ganglion, and the thoracic cardiac nerve, arising from the second thoracic ganglion, and the sympathetic trunk under the cervicothoracic ganglion.

Consequently, other new cardiac nerves, arising from the surrounding sympathetic trunk or ganglion, were found to run along the anomalous left VA without any shift of cardiac nerves as well as the RERSA. Moreover, a ganglion, incompletely differentiated from the cervicothoracic ganglion, was also observed around branching points of additional cardiac nerves along the anomalous left VA. According to Hoffman (1957), numerous ganglion cells are scattered throughout the cervical sympathetic trunk. The general and above cervical ganglions may thus have differentiated from the nearby sympathetic trunk, depending on the appearance of the anomalous great vessels.

In addition, a segmentally-linked accompanying tendency of the left cardiac nerve was recognized: the cardiac nerves accompanying the left VA were more caudally located than the nerves accompanying the left common carotid artery, and

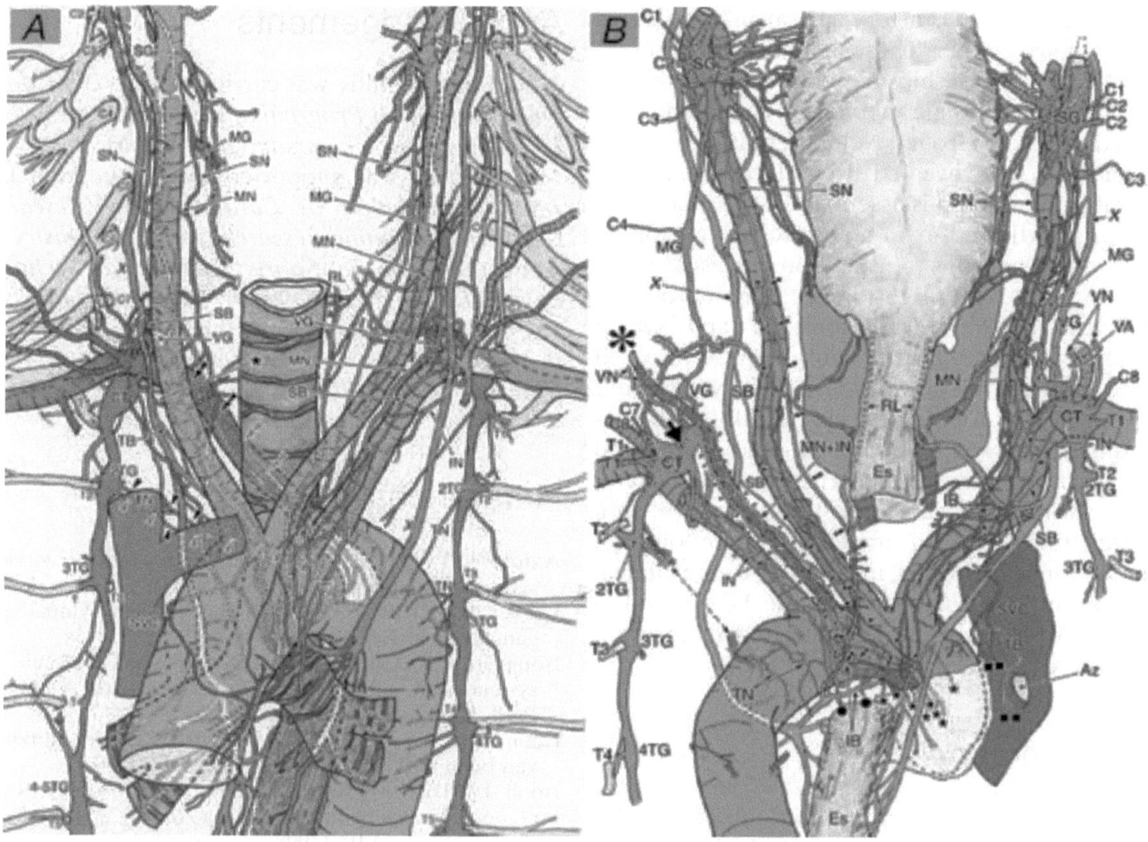

FIG. 7.8. Topological changes of the autonomic cardiac nervous system (ACNS) in human individuals with anomalies of branchial arteries. (**A**) Retroesophageal right subclavian artery (RERSA) viewed from the ventral aspect. Although the right thoracic cardiac nerves, which have never been observed to accompany the normal right subclavian artery, ran along the proximal part of the RERSA (single arrow heads), all the cardiac nerves were recognized. Double arrow heads and black star show the cardiac nerve along the distal part of the RERSA and conventional finding of the corresponding nerve to the recurrent laryngeal nerve. (**B**) Left vertebral artery originating directly from the aortic arch, dorsal view. The single arrow heads, double arrow heads, and arrows show the direct distribution of the superior cardiac nerve to the heart, the superior cardiac nerve via the recurrent laryngeal nerve, and the inferior cardiac nerve along the anomalous left vertebral artery, respectively. These results in two types of branchial arterial anomalies suggest that additional cardiac nerves along the anomalous artery were found in addition to the normal ACNS composition although the ACNS shift have been considered when a great arterial branching anomaly appeared (After Kawashima and Sasaki 2005, 2007). CT, cervicothoracic (stellate) ganglion; Es, esophagus; IB, inferior (vagal) cardiac branch; IN, inferior cervical cardiac nerve; LC, left common carotid artery; LS, left subclavian artery; MG, middle cervical ganglion; MN, middle cardiac nerve; PA, pulmonary artery; PC, pericardial cavity; PVi, inferior pulmonary vein; PVs, superior pulmonary vein; RL, recurrent laryngeal nerve; SB, superior (vagal) cardiac branch; SG, superior cervical ganglion; SN, superior cardiac nerve; SVC, superior vena cava; TB, thoracic (vagal) cardiac branch; TG, thoracic ganglia; TN, thoracic cardiac nerve; VA, vertebral artery; VG, vertebral ganglion; VN, vertebral nerve; *X*, vagus nerve (see Color Plates, Fig. 7.8).

more cranially than the nerves accompanying the left subclavian artery, because there was no crossing of the cardiac nerves. Judging from the characteristics of the accompanying nerve, the anomalous left VA seemed to have a closer relationship with the left subclavian artery than with the left common carotid artery, as in normal VA branching from the normal subclavian artery. This is consistent with recent embryological evidence.

7.2.5.3 Morphological Significance of the Changes of the ACNS

In conclusion, the branching pattern of the great arteries arising from the aortic arch does not cause a shift of the ACNS, and de novo cardiac nerves along the anomalous artery appear in addition to the ordinary ACNS. These anatomical observations of the relationships between the arterial and peripheral autonomic nervous systems do not involve problems from developmental viewpoints. Recently, it was revealed by molecular embryological studies that it is essential for the migration, axonal growth and axon guidance of developing sympathetic neurons to use blood vessels as guidlines to reach final target tissues (Enomoto et al. 2001; Kuratani 2004). Therefore, the gap between anatomical and molecular embryological research is becoming narrower, and it is necessary to further investigate the topological changes in, and relationships between, the arterial and peripheral autonomic nervous systems. This combined examination will be the subject of the next phase in a future study unifying molecular embryological and anatomical findings, in addition to the clinical application in cardiac surgery, the better understanding of experimental results, and their application to cardiac functions.

7.3 Summary

In this chapter, we described the detailed evolutionary morphology of the ACNS, which could not be clarified by recent imaging analyses, using non-human primates and humans. Because the ACNS has been greatly modified in the evolution of primates, it is very intriguing to examine its dynamic changes among primates. Obtained results may also be useful for anthropological and paleontological studies. Multi-disciplinary analyses of the ACNS, the evolutionary comparative morphological results on the ACNS and detailed combined anatomical observations of the relationships between the arterial and peripheral autonomic nervous systems, are compatible with developmental findings. Accordingly, the gap between anatomical and molecular embryological research has begun to be filled, and it is necessary to further investigate the topological changes in, and relationships between, the arterial and peripheral autonomic nervous systems. This integrated investigations will be the subject of the next phase in a future study unifying molecular embryological and anatomical findings, in order to promote the understanding of experimental results and of cardiac functions and the clinical application in cardiac surgery.

Acknowledgements

A part of this study was carried out by *The Cooperation Research Program of the Primate Research Institute, Kyoto University (2004–2006)*.

This study was supported by grants from *the Japan Foundation of Cardiovascular Research* (2003), *the Scientific Research from the Ministry of Education, Culture, Sports Science and Technology (MEXT)* (No. 16790804, 2004–2006 and No. 19790985, 2007–2009), and *the Japan Society for the Promotion of Science (JSPS) core-to-core program HOPE* (2005–2007).

References

Axford MB (1927–28) Some observation on the cervical sympathetic in man. J Anat 62:301–318

Becker RF, Grunt JA (1957) The cervical sympathetic ganglia. Anat Rec 127:1–14

Botar J (1932) Quelques observations sur l'anatomie du systéme sympathique du Gorille. Bull Soc Anthorop Paris 3:168–179 (in French)

Botár J, Becker A (1939) Die Fasernstruktur der Herznerven beim Maki-Affen. Anat Anz 88:494–500

Botár J, Világhi M, Sere D (1950) Die gemischten Rami communicantes beim Affen. Acta Anat 9:235–250

van den Broek AJP (1908) Untersuchungen über den Bau des sympathischen Nervensystems der Säugetiere. Morphol Jahrb 37:202–288

Chiba S, Suzuki T, Kasai T (1981) Two cases of the right subclavian artery as last branch of the aortic arch and review of previously reported cases. Hirosaki Med 33:465–474 (in Japanese with English abstract)

Enomoto H, Crawford PA, Gorodinsky A, Heuckeroth RO, Johnson EM, Milbrandt J (2001) RET signaling is essential for migration, axonal growth and axon guidance of developing sympathetic neurons. Development 128:3963–3974

Fukuyama U (1982) Macroscopic anatomy of the extrinsic nerves in human heart. Acta Anat Nippon 57:357–380

Gomez H (1958) The development of the innervation of the heart in the rat embryo. Anat Rec 130:53–71

Harman NB (1900) The anterior limit of the cervico-thoracic visceral efferent nerves in man. J Anat Physiol 34:359–391

Hirata K (2000) Anatomic variations in Japanese. In: Sato T, Akita K (eds) Skeltons. University of Tokyo Press, Tokyo, pp 5–6 (in Japanese)

Hoffman HH (1957) An analysis of the sympathetic trunk and rami in the cervical and upper thoracic regions in man. Ann Surg 145:94–103

Horiguchi M, Yamada T, Uchiyama Y (1982) A case of retroesophageal right subclavian artery with special reference to the morphology of cardiac nerves. Acta Anat Nippon 57:1–8

Jakubowicz M, Ratajczak W, Nowik M (2002) Aberrant left subclavian artery. Folia Morphol 61:53–56

Jamieson RW, Smith DB, Anson BJ (1952) The cervical sympathetic ganglia- an anatomical study of 100 cervicothoracic dissections. Quart Bull Northwestern Univ Med Sch 26:219–227

Kasai T (1962) Topographic changes of the surrounding structures of the arch of aorta in various anomalies of aorta in man. Acta Anat Nippon 37:275–292

Kawashima T (2005) Autonomic nervous system of the human heart with special reference to the origin, course, and peripheral distribution of the nerve. Anat Embryol 209:425–438

Kawashima T, Sato K (2000) Morphological analysis of sympathetic cardiac nerves in the Japanese monkey and man. Auton Nerv Syst 37:521–531 (in Japanese with English abstract)

Kawashima T, Sasaki H (2005) Topological changes of the human autonomic cardiac nervous system in individuals with a retroesophageal right subclavian artery: two case reports and a brief review. Anat Embryol 210:327–334

Kawashima T, Sasaki H (2007) Morphological comparison of the cardiac autonomic nervous system between normal and abnormal great arterial branching pattern with a brief review of the literature. Auton Neurosci 132:37–43

Kawashima T, Sato K, Sasaki H (2001) An anomalous case of the thoracic cardiac nerve in the Japanese Monkey. Acta Anat Nippon 76:399–406 (in Japanese with English abstract)

Kawashima T, Sato K, Akita K, Sasaki H (2005) Comparative anatomical study of the autonomic cardiac nervous system in Macaque monkeys. J Morphol 266:112–124

Kawashima T, Akita K, Sato K, Sasaki H (2007) Evolutionary and comparative anatomical investigation of the autonomic cardiac nervous system in the African Cercopithecidae. J Morphol 268:771–790

Kitchell RL, Stevens CE, Turbes CC (1957) Cardiac and aortic arch anomalies, hydrocephalus, and other abnormalities in newborn pig. J Am Vet Med Assoc 130: 453–457

Kolesnikov VV (1935) The relation between the truncs and the cardiac branches of the jugular section of the sympathicus and vagus in some apes. Arkh biol Nauk 39:405–419 (in Russ. With English abst.)

Kumaki K (1980) Anatomical practice data. Second department of anatomy, school of medicine. Kanazawa University, Kanazawa (in Japanese)

Kuntz A, Morehouse A (1930) Thoracic cardiac nerves in man. Arch Surg 20:607–613

Kurt MA, Ari I, Ikiz I (1997) A case with coincidence of aberrant right subclavian artery and common origin of the carotid arteries. Ann Anat 179:175–176

Kuratani S (2004) Evolutionary morphology: Bauplan and embryonic development of vertebrates. University of Tokyo Press, Tokyo

Kuratani SC, Miyagawa-Tomita S, Kirby M (1991) Development of cranial nerves in the chick embryo with special reference to the alterations of cardiac branches after ablation of the cardiac neural crest. Anat Embryol 183:501–514

Loukas M, Giannikopoulos P, Fudalej M, Dimopoulos C, Wagner T (2004) Aretroesophageal right subclavian artery originating from the left aortic arch – a case report and review of the literature. Folia Morphol 63:141–145

Mitchell GAG (1953) Anatomy of the autonomic nervous system. ES Livingstone Ltd, London

Mizeres NJ (1955) The course of left cardio-inhibitory fibers in the dog. Am J Anat 96:285–318

Mizeres NJ (1958) The origin and course of the cardioaccerator fiber in the dog. Anat Rec 132:261–278

Nakatani T, Tanaka S, Mizukami S, Okamoto K, Shiraishi Y, Nakamura T (1996) Retroesophageal right subclavian artery originating from the aortic arch distal and dorsal to the left subclavian artery. Ann Anat 178:269–271

Nonidez JF (1939) Studies on the innervation of the heart. Am J Anat 65:361–413

Okishima T, Ohdo S, Takamura K, Hayakawa K (1987) Morphogenesis of an aberrant subclavian artery in rat embryos after materanl treatment of bis-diamine. Jpn J Pediatr 91: 859–869 (in Japanese with English abstract)

Pauza DH, Pauziene N, Tamasauskas KA, Storopus R (1997) Hilum of the heart. Anat Rec 248:322–324

Pauza DH, Skripka V, Pauziene N, Stropus R (2000) Morphology, distribution, and variability of the epicardiac neural ganglionated subplexuses in the human heart. Anat Rec 259:353–382

Pick J (1970) The autonomic nervous system: Morphological, comparative, clinical and surgical aspects. JB Lippincott, Philadelphia, PA

Pick J, Sheehan D (1946) Sympathetic rami in man. J Anat 80:12–20

Riegele HL (1925–26) Die Innervation des Herzmuskels bei Affen. SB Ges Morphol Physiol 37:57–59

Riegele L (1926) Über die Innervation der Halz- und Brucstorgane bei einegem Affen. Z Anat Entwickl 80: 777–858

Roberts LA (1991) Morphological innervation pattern of the developing rabbit heart. Am J Anat 190:370–384

Saccomanno G (1943) The components of the upper thoracic sympathetic nerves. J Comp Neurol 79:355–378

Sharner RF (1930) On the development of the nerves to the mammalian heart. Anat Rec 46:23–39

Sheehan D, Pick J (1943) The rami communicantes in the rhesus monkey. J Anat 77:125–139

Shimada K, Shibata M, Tanaka J, Setoyama S, Suzuki M, Ito J (1997) A rare variation of the right subclavian artery combined with a variation of the right vertebral artery. Showa Med 57:349–353 (in Japanese with English abstract)

Sonntag CF (1922) On the anatomy of the drill (Mandrillus leucophceus). Proc Zool Soc 29:429–453

Sonntag CF (1924) On the anatomy, physiology, and pathology of the orang-outan. Proc Zool Soc 24:349–450

Sumida H (1988) Study of abnormal formation of the aortic arch in rats: by methacrylate casts method and by immunohistochemistry for appearance and distribution of desmin, myosin and fibronectin in the tunica media. Hiroshima J Med Sci 37:19–36

Suzuki T, Chiba S, Takahashi D, Kasai T (1990) Morphological consederations of the bronchial arteries and cardiac nerves, in the so-called Adachi's type-G aortic arch. Hirosaki Med 42:176–182 (in Japanese with English abstract)

Tanaka S (2000) Vagal cardiac branches. In: Sato T, Akita K (eds) Anatomic variations in Japanese. University of Tokyo Press, Tokyo, pp 513–516 (in Japanese)

Tasaka H, Takenaka H, Okamoto N, Onitsuka T, Koga Y, Hamada M (1991) Abnormal development of cardiovascular system in rat embryos treated with bisdiamine. Teratology 43:191–200

Verberne ME, Gittenberger-de Groot AC, Poelmann RE (1998) Lineage and development of the parasympathetic nervous system of the embryonic chick heart. Anat Embryol 198:171–184

Verberne ME, Gittenberger-de Groot AC, Iperen LV, Poelmann RE (2000a) Distribution of antigen epitopes shared by nerves and the myocardium of the embryonic chick heart using different neuronal markers. Anat Rec 260:335–350

Verberne ME, Gittenberger-de Groot AC, Iperen LV, Poelmann RE (2000b) Distribution of different regions of cardiac neural crest in the extrinsic and instrinsic cardiac nervous system. Dev Dyn 217:191–204

Wilson JG, Warkany J (1949) Aortic-arch and cardiac anomalies in the offspring of vitamin A deficient rats. Am J Anat 85:113–155

Wilson JG, Warkany J (1950) Cardiac and aortic arch anomalies in the offspring of vitamin A deficient rats correlated with similar human anomalies. Pediatrics 5:708–725

Wrete M (1959) The anatomy of the sympathetic trunk in man. J Anat 93:448–459

Zuckerman S (1938) Observations on the autonomic nervous system and on vertebral and neural segmentation in monkeys. Trans Zool Soc Lond 23:315–378

Subject Index

Publisher's Erratum to

Hideki Endo, Roland Frey (Eds.)

Anatomical Imaging

Towards a New Morphology

Preface

In Preface, Table 1 was not presented due to the publisher's error. Table 1 is given below. The publisher apologizes for the inconvenience caused.

Table 1. Anatomical imaging — from past to present

time scale	anatomists	technologies	media	dissection
30.000 BC	Paleolithic hunters	bone, ivory carvings	osseous and dental surfaces	evisceration and cutting up of prey animals
13.000 BC	Neolithic hunters	rock and cave paintings	stone and rock surfaces	
5000 BC	Egytian embalmers		papyrus	preparing human corpses for mummification
3000 BC	magicians, shamans and clairvoyants of Sumerian and Babylonian times	writing only writing only	parchment clay tablets	prediction of good and evil by inspection of the viscera of sacrificial animals
2700 BC	Imhotep (Egypt)	writing only		medical (incl. surgical) treatment of live humans
1300 - 1200 BC		writing only	wax tablets	
900-800 BC	Shushruta (India)	writing only		dissection of human corpses by a brushing technique after softening in water
500 - 400 BC		writing only	bamboo tablets	
400 - 300 BC	Aristotle	writing only		dissection of plants and animals
300 - 200 BC	Herophilos of Chalcedon Erasistratos of Iulis Charaka (India)	writing only		dissection of human corpses
200 - 100 BC			parchment	

Table 1. *Continued*

time scale	anatomists	technologies	media	dissection
0 - 100 AD		writing only, manuscript roll		
100 - 200 AD	C Galen	primitive manuscript book (codex), papyrus codex, parchment codex	paper (China)	dissection of animals
500 - 600 AD		wood cut (China)		
900 - 1100	Abu 'Ali al-Husain ibn 'Abd Allah ibn Sina al Qanuni (latin: Avicenna)	printing press (China)		
1100 - 1200			paper (Europe)	
1200 - 1300	Mondino dei Luzzi			dissection of human corpses
1300 - 1400				
1400 - 1500	Leonardo da Vinci, Michelangelo A Achillini, MR Columbo, J Val-verde, W Harvey	drawing, wood cut (Europe), wood engraving, copper plate engraving, printing press (Europe), incunabula		
1500 - 1600	Andreas Vesalius, G Fallopio	printed book, hand-coloured copper or metal plate engravings, mezzotint co-louring, etching,		
1600 - 1700	JB Winslow, BS Albinus, N Steno			
1700 - 1800	A von Haller, P Camper, PFT Meckel, J Hunter, G Cuvier	lithography, chromolithography coloured wax models, wood engraving (xylography), multi-plate colour prin-ting (Le Blont)	limestone	

Table 1. *Continued*

time scale	anatomists	technologies	media	dissection
1800 - 1900	JF Meckel, FGJ Henle, J Hyrtl, NI Pirogoff, H Gray, HWG Waldeyer	black and white photography, X-raying	photographic plate	studying organs of living organisms
1900 - 2000		colour photography, offset printing, plastination of real bodies digital photography (CCD),	autochrome phtotographic plate, photographic film, positive and negative transparency film	
		computed tomography (CT), magnetic resonance imaging (MRI), positron emission tomography (PET), three-dimensional (3D) reconstruction (from pixel to voxel)	CD-ROM, DVD-ROM	

MIX
Papier aus verantwortungsvollen Quellen
Paper from responsible sources
FSC® C105338

If you have any concerns about our products,
you can contact us on
ProductSafety@springernature.com

In case Publisher is established outside the EU,
the EU authorized representative is:
Springer Nature Customer Service Center GmbH
Europaplatz 3, 69115 Heidelberg, Germany

Printed by Libri Plureos GmbH
in Hamburg, Germany